FINE THINGS

Fennel's Journal

No. 8

FINE THINGS

By

Fennel Hudson

2017
FENNEL'S PRIORY LIMITED

Published by Fennel's Priory Limited

www.fennelspriory.com

Limited edition magazine published in 2012
Abridged eBook published in 2013
This extended edition published in 2017

Copyright © Fennel Hudson 2012, 2017

Fennel Hudson has asserted his right under the
Copyright, Designs and Patents Act 1988
to be identified as the author of this work.

All rights reserved. No part of this publication may be reproduced, stored in a retrieval system or transmitted, in any form or by any means, electronic, mechanical, photocopying, recording or otherwise, without the prior permission of Fennel's Priory Limited.

"Stop – Unplug – Escape – Enjoy"
and The Priory Flower logo are registered trademarks.

A CIP catalogue record for this book
is available from the British Library.

ISBN 978-1-909947-30-6

Available to purchase in other formats at
www.fennelspriory.com

Designed and typeset in 12pt Adobe Garamond Premier Pro.
Produced by Fennel's Priory Limited.

> *"To be nobody but yourself in a world which is doing its best, night and day, to make you everybody else means to fight the hardest battle which any human being can fight."*
>
> e.e. cummings

STOP – UNPLUG – ESCAPE – ENJOY

This book, and the series to which it belongs, is about freedom. It's also about the adventures to be had when pursuing one's dreams, developing and communicating one's self, and striving for a slow-paced rural life. It's your opportunity to take time out from the stresses of modern living, to stop the wheels for a while, unplug from the daily grind, escape to a quiet and peaceful place, and enjoy the simple life. Because of this, I'd like you to read it in a distraction-free and relaxing environment: your 'safe place' where you can savour quality time and, if possible, delight in the beauty of the countryside.

That's why the book is pocket-sized, has a waxy cover, and is printed using a special waterproof ink. It's designed to be taken with you on your travels. Don't store it in pristine condition upon a bookshelf; allow it to reflect the adventures you've had. Use a leaf as a bookmark and annotate the pages in the spaces provided with ideas of how you will honour your right to 'never do anything that offends your soul'.

The more mud-splattered, grass-stained, and pencil-scribbled this book becomes, the more you've demonstrated your ability to pursue a contented country life. So go on: live your life, be authentic, and always remember to 'Stop – Unplug – Escape – Enjoy'.

fennel

INTRODUCTION

Individuality, personality, uniqueness, character. They're all words used to describe who we are. They reference one's self. But look up 'self' in a thesaurus and you'll get more words than this. You'll find: originality, idiosyncrasy, eccentricity, peculiarity, mannerism, quirk, foible, idiom, specialty, gift, and distinction. It seems that, depending upon which side of the thesaurus-writer's gaze we sit, one's uniqueness can be deemed to be either eccentric or distinctive. Both, in my opinion, are good. It's better to be individual than a clone of someone else. Worse though is if you don't have a choice in the matter. There might be *expectations* of how you should look and behave.

There's a scientific experiment (the name of which escapes me) where two mice are placed into a box that has a wire mesh floor. The mesh is connected to a battery and, every now and then, an electric current is placed through it. The mice, feeling the shock, jump about and attempt to escape. The current is switched off and the mice return to a restful state. The process is repeated every few minutes. Over time, one mouse jumps progressively harder and more frantically, while the other just sits there and accepts its fate, taking the shocks until it collapses in a catatonic state. It's a test of mental strength and perseverance that applies to us as much as it does to laboratory mice.

Both the thesaurus' descriptions and the above experiment can be related to how we move through life. As we get older, and experience more things, those with strength of character and a sense of purpose will grow stronger and fight harder; those who lack identity and direction might end up sitting in a corner somewhere, blindly taking all the knocks that life throws at them. What does this teach us? That character and purpose are directly linked to confidence and conviction. What links them? Courage – to be oneself, no matter what others might say.

Of course, it's easy to lose sight of the end goal or the person we're trying our best to be. There are a great many pressures and distractions in life. When outside influence is greatest, it's the small, sentimental things close to home that remind us of who we are. They help us to retain our identities amidst the chaos. I call them the *Fine Things* of life.

A book about Fine Things, such as this one, should come with a warning: that the items described herein have the potential to make you very happy, and very skint. It could equally be entitled *How to Max-out your Credit Cards in Twelve Easy Steps*. So read with

CONTENTS

Introduction. 1
The Curse of the August Dumpling 5
Hobnailed 13
This is Me 21
The Reverend James 27
A Rucksack from the Lakes 37
Kelly and Me 43
A Suitable Age for a Beard 49
The Canvas Effect. 59
X Plus One. 67
Botanically Brewed. 75
The Pen that Flew the Atlantic, Twice! 85
The Guv'nor. 93
The Silent Roar 101
Experiences at the Flat Cap Café 105
Four and Twenty Blackbirds. 113
Two Plus Four 121
Tip to Toe 129
Legacy 137

About the Author. 141
The Fennel's Journal Series 145

INTRODUCTION

caution, keep an open mind, and remember that quality of life and one's loving relationships with others are the ultimate Fine Things. That said, there are some 'wonderfully fine' things included, shared because they have brought me pleasure. There's no product placement, advertising or sponsorship going on. The chapter about beards is not funded by Father Christmas, or Zeus, or Captain Birdseye. Fennel's Journal is not 'for sale'. I'm paying homage to my favourite things. Simple as that.

If you are to take but one thing from this book, it is this: be yourself. Be confident enough to show your true self to the world. You might be deemed unfashionable or aloof, but it doesn't matter. Enjoy being you. Have fun. And be different. As Billy Connolly said, "Never trust a man who, when left alone in a room with a tea cosy, doesn't try it on".

Stop – Unplug – Escape – Enjoy

What are your most special, sentimental, Fine Things?

I

THE CURSE OF THE AUGUST DUMPLING

It was not the usual sort of school trip, not even a usual sort of day. But on an August morning during the last decade of the twentieth century, thirty or so spotty teenagers crammed into a rickety old bus and travelled from the urban sprawl of the West Midlands to the free and open spaces of the Derbyshire Dales.

The trip was to be the geography department's first and only field trip of the year: one week's rambling across the hills and dales of the Peak District for pupils to study land and river formations. I was one of the schoolchildren.

Mr Hill, our aptly titled geography master, had told us that the expedition was "scientifically important". In fact, it was so important that we'd been instructed to bring our own toilet paper. (The school governors couldn't have us venturing forth into the wilderness without such luxuries.) Whilst some pupils had skimped on their toiletry supplies in favour of rucksacks full of crisps and fizzy pop, I'd come prepared with four rolls of Ivory White and two rolls of Apricot Peach.

FINE THINGS

I was looking forward to making an impression during the trip. Geography was one of my strongest subjects. But more importantly I wanted to show my townie schoolmates that I was completely comfortable being in the countryside. Whilst they would inevitably be wearing white trainers and T-shirts, I decided to make a statement. I arrived in my battered old walking boots (the ones crusted with muck from the farmyard), my threadbare corduroy trousers, and my wax jacket; such was my resolve to demonstrate my country character. The plan was going well until mid-day, when the heat of the summer sun turned the inside of the bus into a sauna of adolescent perspiration and teenage tantrums. Added to this were the scents of a slowly cooking wax jacket and manure-infused walking boots. So you can imagine who was asked to sit at the front of the bus 'for the good of the class'.

Sitting there next to the teachers, and alongside Travelsick Terry, was bad news for my street-cred. The trip was supposed to be my opportunity to shine, to show everyone how amazing the countryside is and how I come alive amongst the pastures. Fat chance of that happening now. By the time we pulled up at our destination – the lovely and historic market town of Bakewell – the class had already written three verses of a song that ended with me being the town's prize pudding.

The experience, however humiliating, was not

without its merits. This, after all, was an educational trip designed to teach me something. And it did. I'd been ridiculed since noon for my 'inappropriate' dress and for showing my true colours. As a young man charged with finding my independence and individuality, this moment was the turning point. I had two options: I could prove them wrong, lifting my head high and carrying on regardless, or I could grab my loo rolls and do a runner. I did the latter.

I fled the bus, running across the market square towards the first haven I could find: a shop entrance, tucked back from the main road and safely away from the fists of Mongo Chutney, the class Neanderthal. In the window of the shop was a faded cardboard sign. It said: "All Welcome". The invitation was well timed.

I fumbled at the door handle and entered my refuge. Inside, the room was warm and quiet but for the tick-tock of a grandfather clock at the far end of the room. I looked around. The shop was an old-style establishment, with dusty wooden floors and heavy oak panelling supporting rack upon rack of shelving that reached from floor to ceiling.

"I take it your friends won't be joining you?" said a voice from the far corner of the room. I looked up to see an elderly gentleman in a neatly pressed black velvet suit and red bow tie. "Good," he said, "they don't look like my sort anyway. You, on the other hand…"

The elderly man ushered me further into the shop

while asking me why I should be wearing a wax jacket on such a swelteringly hot day.

"I wanted to get straight 'out there' into nature," I replied, "across the fields and up the hills."

"Really?" said the old man, "You looked like you were doing a runner."

"Well, possibly."

"And yet you look so smart, at one with yourself on this fine day."

"Thank you."

"Of course," he said, leaning forward, "there is something missing, something *very important*." He then tapped his head with the index finger of his left hand. "No dress is complete without a suitable titfer."

"A what?" I asked.

"A *Titfer Tat*, a hat," he replied. "And what better place to find one; we're the best milliners in Derbyshire."

The man then held aloft his arms and rotated his wrists like a conductor readying an orchestra. My eyes lit up as I marvelled at the shelves, which were overflowing with hundreds of hats in every conceivable shape and size. From neatly brushed top hats and bowlers, to country caps and deerstalkers, this shop had them all.

"Don't be overwhelmed," said the milliner, "there's a hat in here for you, one that wants to sit upon your head and no one else's."

I looked at the beautiful tweed, felt and moleskin fabrics. There were hats here for every occasion.

"Your hat will call to you," said the man. "Just follow your heart and you will find it. But remember, your head mustn't rule your decision. It's a hat's responsibility to look after your bonce, not the other way round."

The milliner stood back and let me search up and down the racks. I ran my fingers over the headwear, sensing their coarse and the smooth textures, feeling the light and the heavy fabrics. A tweed trilby first took my fancy; it was grey and fawn with a mallard's feather in its band. But it wasn't right for me. A brown bowler looked smart, but sat awkwardly on my head; the panamas were right for the time of year but a little too 'middle wicket' for my liking, and the farmer-style flat caps were too predictable; I had enough of them already.

I was making my way towards a rather splendid-looking deerstalker when I spied, on the bottom shelf, a pile of mixed caps. Their jumble suggested that they were destined either for the dustbin or a local scarecrow. I rummaged through the stack, dismissing the obvious in search of the unique. There! There at the base of the pile, beneath an olive York cap was something different. Something unique. Something I just had to try on.

I prised the hat from the pile and carried it to a mirror on the far wall. I placed it onto my head. The fit was perfect: comfortable and comforting. It felt right, as if I should hold it to my cheek rather than place it over my brow. But there was a problem. If love is blind, then teenage love is like sketching with charcoal at night.

This was the largest, podgiest, floppiest cap I'd ever seen. Eight segments of tan and orange-coloured tweed, bunched together beneath a small button at the crown, yet saggy enough so that it flopped over my ears and bulged like a puffed cushion atop my head. It looked like an oversized Yorkshire pudding.

Fate had brought me here. I'd followed my heart to this hat. But was destiny meant to look like this?

"Voila!" said the milliner in a loud cheery voice. "The hat maketh the man!"

"But it's so…big!" I stated, as I lifted up the bulging fabric so that I could see.

"Indeed it is," he replied, "but have you ever seen anything so splendid?"

"Or so…ginger!" I exclaimed.

The milliner was right, of course. I was sixteen years old and only 4ft 10in tall. The hat added much needed stature to my punitive frame, making my head appear twice its normal size and adding five inches to my height. It would certainly get me noticed and prove that I was unaffected by the ridicule of others. Keen to demonstrate my newfound confidence, I decided to purchase this ridiculously magnificent hat.

"How much, dear sir, for this excellent hat?" I enquired.

"For craftsmanship of this nature," he replied, "fifty pounds is the price."

My heart sank. I had only thirty pounds – my entire spending money for the week. I took it from my pocket and counted it in hope that it had magically grown during my time on the bus. I glanced towards the shop window and to the street beyond. There were my schoolmates clambering back into the bus, and my teacher looking anxious.

Sensing my disappointment, the milliner approached me, straightened the hat on my head, and smiled. "Of course," he said, "I could accept thirty."

I gasped with excitement and thanked the milliner. I handed over my money and walked, with the hat on my head, out of the shop. The hat bounced, wobbled and nodded with approval. It had, at last, found a loving owner.

The lads on the bus were less appreciative.

"Hey everyone, there's the pudding, and he's wearing an August Dumpling!"

It mattered not. I held my head high, climbed aboard the bus, sat down, and ignored their taunts. I didn't care what they'd said. This was *my* hat, and one like no other. Plus, contrary to their intentions, I liked the name that these bullies had given it. The August Dumpling was born. It would become my all-time favourite hat, and the origin of my love of Fine Things.

Stop – Unplug – Escape – Enjoy

What one thing most defines you?

JANUARY

II

HOBNAILED

When I was younger, I had a utopian dream of life without change or compromise. One where I'd never have to do anything I didn't want to. I'd be able to sit on the garden swing all day, sucking sherbet through a liquorice straw and wondering what to do with a never-ending supply of pocket money. And then things changed. The rope broke on the garden swing, my Dad stopped my allowance, and my once-loved confectionary manufacturer blocked up the hole in the liquorice straws. Change, it seems, is inevitable. And so is compromise.

Things today 'sure ain't like they used to be'. They've changed, moved on, *progressed*. These days we can eat our fish 'n' chips, real or metaphorical, from a wrapper that's printed to look exactly like newspaper. And while we're doing so we can read headlines such as: "Mr Chippy's Chippiest Chip Gets Even Chippier!" Informative, nutritional (well, nearly), and oh so 'authentic', the new version is so much better than the original. Or is it?

'In with the new, out with the old.' That's what we're told. We have to change, adapt, *compromise*.

But what happens when things move too quickly for us to adapt or, heaven forbid, we don't value the new item? "What was wrong with the old one, or the way we used to do it? Why is this sachet of Instant Black Tea any better, or more instant, than a spoonful of tea leaves?" Change for the sake of change is not good. The compromise is too great.

I've always believed that a person has to be comfortable in his or her own skin. There's no point pretending to be someone or something we're not. We should be authentic: the 'real deal'. *Neither a clone nor mimic be.*

I resisted change for as long as I could, refusing to accommodate other people's expectations until I absolutely had to conform. This occurred shortly after leaving college when, during my first winter working full-time as a gardener, frozen ground put paid to my earnings. I had to retreat indoors and wait until there was something to do. Living off my overdraft, and sinking into more than debt as I battled the long dark evenings, I realised that my horticultural dream was not to be. I would have to seek a 'proper job', one that, if it were to provide year-round security, would mean working *indoors*. I shuddered, feeling as cold as the snowdrops that were refusing to show their heads in my garden.

And then it hit me: I would have to attend *an interview.*

The problem with interviews is that they're the

civilian equivalent of a Firing Squad. (When else are we asked to sit bolt upright, on our own, opposite three unsmiling people, and forced to defend our rationale for studying G.C.S.E Needlework?) In any other meeting, we'd at least be allowed to blink without thinking it might betray some inner weakness.

"*Did you see how his eyes moved when we asked him about his career aspirations?*" they might say.

"*No, but I noticed he drank our complimentary glass of water using his right hand for the first half of the interview, and then switched to using his left hand towards the end.*"

"*He's shifty. We don't trust him. Next candidate!*"

I'd never attended a formal job interview, but I knew of its reputation. So I read up about how best to behave in the meeting. I learned all I needed to know. There were five golden rules:

1) Always mirror the interviewer's movements: if they smile, you smile. If they lean forward, you lean forward. *If they break wind, do I do the same?*

2) Look interested. Make lots of eye contact. And remember, they will be reading your eyes.

3) Have your answers prepared, and know what questions you want to ask.

4) Be smartly and professionally dressed.

5) Above all, make a *positive lasting impression.*

Encouraged by my research, I responded to a newspaper advert for a Shop Supervisor at my local garden centre. "Should be a breeze," I thought, "packets of seeds and a few plastic plant pots won't take much supervising; so I'll give it a go." The interview date came through. I had one week to prepare. I ran through my preparation notes. All understood, with the exception of "be smartly and professionally dressed". What did that mean? I wasn't going for an office job, but then the advert did say 'Supervisor', so I'd need to look respectable. Supervisory. *Authoritative.*

One of my 'unbreakable' rules about compromise concerned my choice of clothing. It stipulated: "If I can't garden in it, then I won't wear it". Consequently, my wardrobe was rather limited. I had collarless 'grandad' shirts, corduroy trousers, woolly socks and stout undies, sturdy leather boots, baggy jumpers, a leather jerkin, several hats, and a ten-year-old wax jacket. That's all. I definitely didn't own a suit, tie, or shoe polish. If I changed my wardrobe I'd be robbing my 'self' of who I am. But I'd also be risking my chances of employment. So I went shopping – to the place where I'd purchased all my clothes since art school: the local charity shop. There I purchased a navy blue blazer, a stiff-collared lilac shirt, a red-and-black Paisley tie, a pair of herringbone tweed trousers and some white patent-leather slip-on shoes. "Very executive," I thought. "Just what I need to make a positive lasting impression."

I returned home and tried on the clothes. They were, shall we say, 'challenging'. The shoes were four sizes too big, the tie was six inches wide, the jacket stank of mothballs, and the trousers were so tight they must have been made for the soprano in a male voice choir. But more worryingly, this new look wasn't me. As I stared into the mirror all I could see was the awkwardness of compromise. I felt completely shoehorned into being something, or someone, I was not. But I reasoned that any feeling of discomfort could only be compared to the 'breaking in' process for a new pair of walking boots. I just hoped that the blisters wouldn't be too agonising and the terrain not too tough. Things, I naively assumed, could only get better.

A square peg has to be whacked very hard for it to fit into a round hole (ever wondered why certain situations give you a headache?), so it pays to know what we're getting ourselves into. Leaps of faith tend to favour those with long legs, so I rang the garden centre and asked them to confirm the dress code for the interview. "Business casual," they replied. *Business casual? What did that mean? My research didn't tell me about that.* "Jacket and shirt, tie optional," they clarified. Ah, good, it wasn't as bad as I'd feared. I dumped the shoes and tweed nutcrackers in the bin and put on my best corduroy high-rise trousers (the ones with the Albert Thurston tweed braces), and my new William Lennon hobnail boots. They didn't quite match the blazer, shirt and tie, but I figured the interviewers wouldn't be looking under the table. "This I can live with," I thought, as I tightened my tie and stood,

James Bond-like, in the mirror. "Ready for action, albeit a little *shaken not stirred*."

The interview day arrived. I travelled by bus to the garden centre and stood outside the manager's office waiting to be seen. I was invited in and offered a chair. I sat down. The interview began.

"Thank you for attending," said the interviewer. He leaned forward. I leaned forward. He smiled. I smiled. He rubbed his nose. I rubbed my nose. He raised an eyebrow. I raised an eyebrow. He frowned. I frowned. He shuffled in his chair. I shuffled in my chair.

"Let's begin," he said. "Can you tell me about yoursel-"

"I want this!" I screamed. "I want it so much. I'll prove it to you. I'll talk to you with my eyes. I'll show you my experience. I'll make it memorable for you. I'll give you what you want. I'll bend over backwards, anticipating your every move. I'll work my knuckles to the bone..."

"Oooo-kaaay. Anything else?" he replied.

"Give it to me! You know you want me, so show me the package."

The manager paused, looked me up and down, and said, "Let's talk about the image".

"What image?" I asked.

"Well, we couldn't possibly have you looking like that."

"Like what?"

"Like a seventies game show host who's moonlighting as a chimneysweep."

"Pardon?"

"Mr Hudson, you have to understand that we have standards at this company, and expectations for whomever fills the role. We have our *corporate brand* to uphold. Our Shop Supervisor is required to lead by example." He paused, and then said, "Would it help if we provided a uniform?"

"If that's what it takes for me to get the job, then yes," I said.

"Good. Then we have a deal. You may start on Monday."

That's how I ended up wearing company issue pea-green trousers, a yellow shirt, an orange jumper and a badge saying "Here to Serve". And in doing so, I lost a part of me that was previously unbreakable. I knew that I was stuck, wedged into a square hole.

I looked at my favourite boots and said, "My friends, I've been well and truly hobnailed".

Stop – Unplug – Escape – Enjoy

Which of your values are absolute, that you would never compromise?

JANUARY

III

THIS IS ME

> *"It takes courage to grow up
> and become who you really are."*
>
> e.e. cummings

Some things, sadly, do not change. Like the compromises we make for others. Dancing to their beat, we wonder when the music will stop. But compromise is a choice. As is the defence of one's self.

I had a life-defining 'choice' moment today. You know, the sort where you can either passively or passionately react. I was at work and my manager called me into her office to review my annual performance. I sat down at the opposite end of her desk, placed my hands on my knees, took a deep breath, and stayed silent.

"Fennel," she said, as she frowned and leaned forward, "before I give you your grade for the year, I need to talk to you about your behaviour at work. Whilst I'm impressed by your leadership skills, energy, and creativity, you have a tendency to be 'unpredictable'. It's as if you seek to do your own thing regardless of the line I expect you to tow. This makes it difficult for me to know how you'll respond to my requests, and therefore I'm challenged by how to motivate you. It's as if you arrive at work with a bunch of daffodils in one hand and a chainsaw in the other."

"You're *completely* unpredictable;" she continued, "it seems that you're happy to contribute ideas and lift the spirits of the team, but if I insist upon you cancelling your holiday to prove to me where your allegiance lies then you become all defensive and uppity." She sat back in her chair, folded her arms, and said, "I'd like you to complete a personality test".

There are a number of these tests available. Each attempting to help the individual to better understand their preferences and likely behaviour. (And, of course, for their manager to understand what strings to pull to get them to perform.) I'd done them before, and they usually resulted in me being declared either an 'introverted, intuitive, feeling, perceiving' type or a 'shaping, implementing, plant'. This apparently means that I'm a visionary who likes to be the one driving things, can make things happen but has a tendency to focus on strategic thinking and idea generation rather than completion and delivery of the task. Knowing what the results of yet another test would be, I decided to answer it 'differently'.

A week later, and back in my manager's office, I had the joy of 'explaining myself' to my boss and a representative from Human Resources.

"We have a challenge," said my manager as she glanced at the HR person for support. "The results of your assessment were…inconclusive. Your report shows that you're extrovert but don't like time with others, thinks about things and then makes a decision based upon intuition, senses what's right and then gets emotional about it, is widely perceptive but highly judgemental, is very inventive but relies upon others for ideas, and enjoys finishing tasks even though you choose not to."

"It seems," said the HR person, "That you're rather… *individual*."

Relaxing in my chair, I smiled knowing that the words were the best compliment I could have been given.

The reason I'm telling you this is because, after the meeting, I was handed the results of the test. At the bottom of the page was an inspiring comment about personality traits. It said: *"An individual's default personality type is established by the age of fourteen, after which it is virtually impossible to change. While experience and education can influence behaviour, a person's default preferences will always be the same."* This made me think of the August Dumpling and how – at a crucially young age – it helped me to know and be myself.

Maybe, if I hadn't discovered the milliner's in Derbyshire, I would have become a different person? Apparently not. I was sixteen; the die was already set.

Robert Louis Stevenson said, "To be what we are, and to become what we are capable of becoming, is the only end of life". So, contra to the small print on the test results, I believe that we can always choose who we are and what we'll become. Whilst our 'child within' might have the same default personality, we can adapt to our environments and life goals.

Knowing one's personality and motivators helps us to understand why certain things are special, and why some Fine Things appeal to us and not others. They're the emotional purse that attributes value to things.

If you were asked to write a statement explaining 'this is me', what would you write? Would you define yourself as "a Shaping Plant who uses Intuition and Feeling to Implement things"? Of course you wouldn't. You know yourself better than that. You'd describe yourself by the values, labels or actions that define you.

I've just completed the task, writing down that I'm individual and fiercely independent, an outdoorsman who seeks freedom in quiet places amongst nature, a traditionalist who loves old-fashioned and handcrafted things, a lifestyle author whose motto is 'Stop – Unplug – Escape – Enjoy', and a husband, father, and friend who's there for others.

Interesting how the list developed, isn't it? First I wrote what defines me, regardless of others, and then I wrote how others influence my perception of 'me'. Perhaps that's the mark of an introvert and that an extrovert might lead with how others make them who they are? But it's a personal view. Others might describe us differently, based upon what they observe.

Does your behaviour, appearance, words, and actions, reveal the real you? Are they congruent with your sense of self? Sometimes we have to add a simple statement to aid others' interpretation of us, so that our actions and preferences are explained. Just like I did earlier. Failure to do so, or having an incongruent display of self, can confuse others.

I'd confused my manager. It wasn't her fault that I'd hidden my true self at work; that I was finding the fast-paced and competitive pressures of business a stressful contrast to the quiet times in the countryside that I so needed and love; that the 'thick skin' I was expected to wear at work was weighing heavily upon the sensitive nature of my soul; that the loyalty and devotion I was expected to give to the company was at odds with my desire to be free; and that 'following suit' with corporate image and culture was suffocating my principal desire to individual and independent. I was quite simply doing the wrong job for the wrong company.

Proud of my self, and forever at odds with those who try to imprison or control me, I was the mouse who refused to lie down.

Stop – Unplug – Escape – Enjoy

How would others describe you?
Does it match how you see yourself?

February

IV

THE REVEREND JAMES

This morning I discovered that my work shoes had finally succumbed to their life on the hamster wheel. A hole the size of a conker had formed in the sole of my left shoe. It gaped and threatened to eat up every bit of grit and swallow every puddle I happened to stand in. The situation was unacceptable. I'd not long purchased the shoes. In a flabbergasted state, I asked Mrs H if she knew why they'd perished so quickly.

"Not long had them?" she replied. "You bought them before we were together. Eight years ago! That's a respectable lifetime for a pair of shoes."

"No it isn't," I said, "not by a long way. I've a pair of walking boots that are twenty years old and still going strong."

"You can hardly compare a commando rubber sole with one made of leather."

"Or maybe it's because my boots were made for the job and these work shoes were designed for people who spend more time sitting down that they do standing up?"

"Fennel. You have an office job. You spend all

day sitting on your backside. The reason your shoes have worn out is because you drag your feet on the way to work."

"That's true. But I walk with a spring in my step when I'm coming home."

"You're right. But why are you so emotional about a pair of shoes? You can replace them easily enough, and besides, it's Saturday morning. You've plenty of time to get a new pair."

"I don't want to replace them. I like the thought of them wearing thin; it helps to remind me that they, like my work situation, are not permanent."

"Oh no, here we go. You're about to have one of your anti-work rants again. The *woe is me, the disgruntled employee, the internal conflict, the suffering poet…*"

"I'm not. I'm learning to live with all that. It's just that I'm bored. I need to do something exciting, something adventurous. I need to be somewhere *inspiring*."

"Oh, I get it now. I forgot. It's February; you've got your winter blues."

"I have. I've been cooped up for too long. I feel like I've been buried underground with a gone-off bottle of banana liqueur."

"Thanks a lot."

"You know what I mean. I need space and something to renew my appetite for life."

"You need to be outdoors. Away from here. You need a *holiday*."

"We both need a holiday, but I have commitments. It will be months before we can go anywhere."

"Not so. It's the weekend. We have plenty of time. Get your coat; we're going for a drive."

That's how I found myself sitting in the passenger seat of Mrs H's car, with her at the wheel and Little Lady Hudson sitting contentedly in her baby seat in the back.

"You said you need to be somewhere inspirational," said Mrs H, "so we're going to follow in the footsteps of St. David and honour his teachings to 'be joyful, to keep the faith and to do the little things'. We're going to Pembrokeshire, to Solva and St. David's, where we can walk along the coastal path and spot the terns and choughs that nest upon the cliffs. We'll experience the 'infinite freedom' that can only be experienced beside the sea. And, if the weather's harsh when we arrive, there's a little pub where you can drown your sorrows. You and I both love Wales, and Pembrokeshire is one of its most beautiful counties. This is going to be a pilgrimage-and-a-half."

We travelled quickly from Oxfordshire, cross country to the M4 motorway, over the Severn Bridge and into Wales. We passed Cardiff, Swansea, Carmarthen and Haverford West until we reached the western-most point of the country. And the sea view when we got there? There wasn't one. A dense sea fog had drifted ashore, obscuring the normally spectacular coastline and forcing Mrs H to drive slowly to our destination: the harbour village of Solva.

It was lunchtime when we arrived, yet the village looked deserted. Mrs H parked the car; I opened the door and stepped outside. It was eerily quiet. The fog was icy cold and smelled of seaweed. There was no questioning that we were near to the coast.

"The sea's in that direction," said Mrs H as she pointed into the mist. "On a clear day you can see moored boats, lime kilns and cliffs, the coastal path, and at the far end is a delightful cottage with a terraced garden that overlooks the harbour. But not today. Today I think we're going straight to the pub."

Standing next to the car, looking into peaceful nothingness, and feeling the dampness and chill of the winter fog, was like being infinitely shrouded from my troubles. I had my family with me, and yet felt wonderfully and enticingly alone. My mind was free to drift into the mist and return when it needed to settle. I could see and hear so little, and yet felt so much. It was only four hours since I'd complained about being trapped by an indoor life, yet now I was filled with love for the world. I was reminded of Thoreau's words: "One must be out-of-doors enough to get experience of wholesome reality, as a ballast to thought and sentiment. Health requires this relaxation, this aimless life".

I could have stood beside the car and done nothing

else. If I had, I would have returned home a happy man, knowing that I had settled my nerves. But Mrs H's mention of the pub got me yearning for a pint of ale and the warmth of an open fire.

(The Greek author Herodotus wrote in 420BC, "If a man insisted always on being serious, and never allowed himself a bit of fun and relaxation, he would go mad or become unstable without knowing it". Which, as I see it, is good enough reason for a pint or beer.)

"It's not far," said Mrs H as she took my hand, "just over there through the fog. If you think you're happy now, wait until you get inside."

Mrs H collected Little Lady from the car seat and the three of us made our way to the pub. We entered and immediately felt the warmth of a log fire. The bar was to the left, but was an adults-only area. So Mrs H and Little Lady went into the lounge to the right while I went into the bar to get the drinks. (A man walks into a bar. "Ouch!" he says.)

The bar in the Harbour Inn at Solva is exactly what you'd hope for in a seaside pub. The room is small enough to feel cosy, has a log fire that burns almost continuously, and has a serving counter that's long enough to allow people to congregate and converse. That said, when I walked in, there was only the innkeeper there.

"Fog's kept them away," said the innkeeper (who, for the record, was sporting a rather impressive handlebar moustache and tartan waistcoat). "It's usually busier than this. I was wondering if anyone would come in."

"It's not kept me and my family away," I said. "We've travelled two hundred miles for this."

"Then you'll be needing a drink. What shall it be?"

"What's your best real ale?"

"You'll be wanting a pint of the Reverend."

"The what?"

"The Reverend James. It's Brain's best. Brewed to an original recipe from 1885 and named after the Revered James Buckley of the Crown Buckley brewery. It's just £3.75 a pint."

"Good value; but a vicar who owned a brewery? Sounds a bit far-fetched."

"Not in these parts. And besides, you gotta think about it. It's a supply and demand thing. If you want a long line of people seeking redemption, you might as well help them to sin."

"There's logic in that. I'll try a pint of this demon drink, in a glass with a handle, but spare the guilt."

The innkeeper reached above his head and grabbed a pint mug, then levered the pump in front of him until the glass was full with dark ruby ale that had a yeasty, frothy, head.

"Here, try that," he said.

I lifted the glass to my mouth. The beer had a sweet, toffee-like aroma. *Mmm. This is going to be good.* I took a sip. The ale had a malty, almost nutty, flavour with a berry-like richness. It reminded me of my favourite ales: Adnams *Broadside*, Wychwood *Hobgoblin* and Brakspear *Triple*. But it had something else: a 'roundness' of flavour that comes from well-kept cask ale, which can never be replicated in the bottled or kegged version.

"This is the best ale I've ever tasted," I announced.

"Thought it might be," replied the innkeeper. "Worth the trip?"

"D'you know, I think it is."

"So, you're here on holiday?"

"No," I replied, "just a day trip to clear the cobwebs and, by the name of this beer, to partake in a sermon of sorts."

"Then you'll be seeking enlightenment," he said. "How about, while you drink your beer, I tell you some interesting facts that you can then share with your wife next door?"

"Okay, hit me with them."

"First off: ever been asked to mind your P's and Q's? Well, you know that ale is served in pints? In olden days it was served in pints and quarts. When customers became too rowdy, the bartender would tell them to settle down and mind their own pints and quarts. Hence the expression: 'Mind your P's and Q's'."

"That's amazing!"

"Secondly, did you know that before thermometers were invented, brewers would dip a thumb or finger into the mixture to check the temperature was right before adding the yeast. Too cold and the yeast wouldn't grow.

Too hot and it would die. Dipping one's thumb into the brew is where we get the phrase 'rule of thumb'."

"Wow!"

"Thirdly, did you know that in olden days, frequenters of pubs drank from ceramic mugs? These mugs had a little hole made into their rim, so that when the drinker needed a top-up, he would attract the attention of the barman by blowing through the hole, which produce a loud whistle. Hence the expression, 'to wet your whistle'."

"You're a font of knowledge!"

"And how about this one: I can tell from your wedding ring that you're a married man. Did you enjoy your honeymoon?"

"It was very pleasant, thank you."

"Drink much beer?"

"Not really."

"Well, did you know that in ancient Babylonia it was traditional for the father of the bride to supply his new son-in-law with all the mead he could drink for one month after the wedding? Mead is made from honey, and as the Babylonian calendar was lunar-based, the period became known as the 'honey moon'."

"A whole month of free beer? I feel short-changed!"

"Oh well. You look happy enough now. Best you get back to your family. But before you go, I've a joke for you."

"Okay, then, tell it to me."

"A man walks into a bar and tells the barman to pour ten pints of ale and line them up as quickly as possible. The barman does as requested, pouring one glass after another and placing them on top of the bar. The man, who has a desperate look in his eyes, begins drinking the beers nearly as quickly as the barman can pour them. The barman says, 'Hey fella, what's the rush?' to which the man says, 'If you had what I've got, you'd be doing the same'. The barman looks at the man and says, 'I'm so sorry; I wasn't meaning to be insensitive. But, if it's not too intrusive, do you mind if I ask what you've got?' The man wipes his lips and says '85p'."

Stop – Unplug – Escape – Enjoy

Where's your favourite place? How does it make you feel, and why?

March

V

A RUCKSACK FROM THE LAKES

*"Take, if you must, this little bag of dreams,
Unloose the cord, and they will wrap you round."*

W.B. Yeats

Returning to work after my time in Wales was like trying to climb a spate-enraged waterfall. I was thrust back into the corporate world with the dignity of a road-killed swallow. Soon I was drowning in emails and telephone calls and being confronted with a gut-wrenching reality: that, at work, I am little more than putty being squeezed through the fingers of those who live to manipulate others. It felt like someone, somewhere, was wringing out a blonde-haired voodoo doll they'd found floating in a pint of beer. I placed my hands on my desk, closed my eyes, and decided that something had to be done.

Monday mornings at work are never easy. (The afternoons aren't that better, either.) There's just too much contrast between the joys of the weekend and the chores we often have to do during the week. Thankfully, we have the restful pleasure of weekday evenings to keep us sane. (Workdays are, I imagine, rather like learning to ice-skate Torvill and Dean's *The Bolero*. They start and end easily enough; it's the bit in the middle that causes the pain in the arse.) I'd not adequately braced myself for the onslaught of work. I wasn't prepared for the 'ragdoll shaking' of normality. But, after Solva and the 'sermon' from the innkeeper, I wouldn't allow the shock of a workday to sink my spirit. I decided to focus on the good things: the things that remind us of who we are.

Whilst some items have sentimental value, there are others that are super-special. They are the Finest Things, totems that not only reflect our personality but encapsulate the spirit of that which inspires us. Picturing or holding them reassures us of who we are and where we want to be. And, lifeline that they are, they help us to endure a crappy Monday.

I have several Finest Things: such as a Conway Stewart 'Wordsworth' fountain pen that I use for book signings; the August Dumpling cap; several 'Priory Red' waistcoats; a Mount Royal pocket watch that I bought with my first wages, which reminds me that time is in my hands; and a box file containing my earliest writing. But there's one item with more power than most. It's a rucksack that holds my leather writing folio, my camera, my binoculars, my flask, my nature notebooks, and my soul. It's the one I take with me on my walks, my adventures, and my dreams. I call it my 'adventure bag', because it calls to me – urging me to leave the stuffiness of my study in search of places where I can feel the earth and touch the sky.

Whilst I've often thought of myself as being hobbit-like, often wanting to bury myself away in a cosy nook under a hill to weather out the storm of life, I know that I'm not one of the "plain quiet folk who have no use for adventures". I yearn for adventure, but lack the absolute freedom to make it an everyday thing. The rucksack is far braver than me, more restless and heroic, as if it's already visited my dreamed-of places and has the map. I hold onto it tightly, knowing that it will carry me there in the end.

My adventure bag has personality. Although I've given it a name, its proper name is Matthew. Named after the outdoorsman and adventurer Matthew Entwistle, who wrote the biography of Lake District adventurer Millican Dalton, it's made by a company called Millican – again named after the Lake's famous adventurist.

The bag's full name is Matthew the Daypack. I love it dearly, as I do its makers Nicky and Jorrit who live in the Lake District. Most admirably, they gave up their helter-skelter city life, went travelling, then came back with a desire to create something 'ethical, sustainable and with a conscience'. Matthew the Daypack, and a bundle of other bags and accessories, was the result.

Like Millican, Nicky and Jorrit looked to the Lake District as somewhere to rejuvenate and inspire them. This makes them valued friends of mine, as both the Lake District and Millican Dalton's lifestyle inspired me at the time I most needed it. Which gives my adventure bag extra meaning. It encapsulates the spirit of adventure and travel, that we're heading somewhere for a reason, to discover something that changes us for the better.

Of course, I could attempt to be unemotional about the bag, telling you that its design was based upon a classic alpine backpack; how it's constructed from heavy-duty organically-grown canvas; has vegetable-tanned leather trims and antiqued brass press-studs, and how it is unique with its front-opening zip and pockets designed to house Ordnance Survey maps. But as soon as I attempt to mention its colour (which I can only describe as 'a sepia photograph warmed by the glow of an autumn sunset') I am immediately reminded of Victorian images of the Lakeland Fells and the words of Alfred Wainwright saying, "I was totally transfixed, unable to believe my eyes. I had never seen anything like this…here was no painted canvas; this was real. This was truth".

C.S. Lewis said, "Even in literature and art, no man who bothers about originality will ever be original: whereas if you simply try to tell the truth (without caring twopence how often it has been told before) you will, nine times out of ten, become original without ever having noticed it". Good words. Which explains why I can't be unemotional about my adventure bag. Its very purpose, its inner truth, is to incite emotion. I love it, and with that truth I can honestly say that its originality is without peer. It's the best Adventuring Matthew Daypack Bag I could wish for.

The daypack represents a promise of adventure, no matter what the circumstance – that I can go there and back in a day. Which is why, in my mind's eye, I'm writing this sitting beside a stream that runs along the base of Honister Pass in the Lake District. I'm looking up at the scree-fall and cliffs in the mountains and can hear nothing but the trickling of water and the bleating of distant sheep. My adventure bag is next to me, absorbing – and helping to create – the atmosphere. I am a free man, here among the mountains and away from the ordeal of a Monday morning. But really I'm still at work, sitting at my desk and keeping my head low to avoid the bullets. Which has enabled me to get a good look at the photograph on my desk. It's a picture of me with the bag, sitting in the location I've just described, and looking so very, totally, contented.

My body might be indoors, being forced towards the frontline of battle, but my mind is in the lakes, with the adventure bag of my dreams.

Stop – Unplug – Escape – Enjoy

What things encourage a sense
of adventure and freedom in you?
How often do you grant them
their wish to explore?

April

VI

KELLY AND ME

I'm going on a picnic with my friend Kelly. We'd like you to join us. It's a fine April day, the breeze is warm and buttercups are in flower. We'll roll out a rug, open a hamper, lie back and enjoy the private pleasures of a spring meadow. And then, if it's your kind of thing, I'll ask you to turn towards Kelly, look longingly at her, and blow gently up her bottom.

Before you accuse me of taking our friendship too far, I'll explain that this is a safe but temporary offer. Kelly will open up to you, I promise, and although she takes some stick for my pleasure, she'll never say a word about our steamy liaison. So while Kelly's keen, I suggest you accept the offer.

Some might say that Kelly has a harsh exterior. True, she's sometimes fiery and hot to handle. But I know what gets her going. She craves wood. The harder the better. But if your wood doesn't last long enough, she'll smoulder with disinterest. So keep thrusting it in and watch her temperature rise. Take a risk and she'll pop your cork. And when she does she'll gargle and spit out whatever's inside her, then light up and smoke like a chimney. And when you've finished dunking your bag, you can replace her top, look down, and have a quick slurp. Who would have thought that the art of tea-making could be such unadulterated filth?

Kelly, in case you were getting overly hopeful, is not a person. She's a kettle. A very special and unique type of camping kettle that uses heat from a fire built inside it to boil water. So while we recover from our earlier 'strain' of tea-making, I'll tell you about this wonderful device.

A Kelly Kettle is an Irish creation, designed to boil water quickly outdoors. It was invented in the 1890s by Patrick Kelly who lived in a small farm on the edge of the shores of Lough Conn, County Mayo. Mr Kelly used his invention to boil water from the lough when fishing for trout and salmon. The original model was made from tin, but was soon replaced by ones made from copper (and latterly, from aluminium and stainless steel). The kettle grew in popularity as more anglers, especially those from the UK, visited the Lough and saw smoke rising from the shore. (Tradition dictated that each angling guide working the Lough would take two anglers out in a boat. At lunchtime, they would row to the shore for lunch, where both the guide and his guests would collect driftwood before returning to their lunchtime spot where the guide would load a Kelly Kettle, light the fire, and make tea. In return, the anglers would provide the guide's lunch and offer him a drop of whiskey. This is still the tradition, with a hip flask never being far away from a Kelly Kettle.) Today, the Kelly Kettle is used around the world. But it's still made by the Kelly family who, I'm pleased to say, continue their boating and ghillie service on Lough Conn. And while other manufacturers attempt to copy the design, the authentic Kelly Kettle is still the best.

My Kelly Kettle is made from aluminium and has a double-walled cylinder with a chimney through its centre. Water is poured into the cylinder, which is then placed onto a saucepan-like firebase that's been loaded with tinder and kindling. Combustible material (such as sticks, twigs, pinecones, dry grass, peat, heather etc.) is dropped down the chimney until it's relatively full, then the kindling is lit through a hole in the firebase. Because the hole in the firebase is smaller than the diameter of the chimney, air is drawn up, creating a furnace inside the kettle. This creates great heat that, if you've loaded the kettle properly, boils the water in less than five minutes.

Using a Kelly is easy, but it's also an art. The challenge is to boil the kettle as quickly as possible, which leads to all sorts of 'black art' techniques deployed by experienced Kelly handlers. My preference is to place two sheets of scrunched-up newspaper into the firebase and one up the chimney. I'll then drop a handful of fine twigs (about a quarter of an inch diameter and four inches long) down the chimney followed by twigs of pencil length and thickness. I'll angle the hole in the firebase into the wind and then light the paper. Blowing gently into the hole will help until the paper burns and the twigs drop down from the chimney; I then add more pencil-thickness twigs into the chimney until the fire can be seen licking out of the chimney top. I'll then either leave the kettle to boil or add some thicker sticks if I intend to keep the fire going afterwards to cook some food.

There are super-secretive types (you know, the ones who had recipes at school for making their conkers harder than all the rest) who own a Kelly Kettle and have somehow mastered the art of boiling it quickly. I've seen them use all manner of tricks, such as burning Mexican Maya fatwood tindersticks (high in flammable resins), pistachio shells and pine cones soaked in paraffin, matches tied into bundles, egg cartons torn into shreds and dipped in candlewax, cotton wool rubbed in Vaseline, and waxed milk cartons cut into half-inch strips. But the ultimate, as seen on the banks of a lake in Devon, was a 'fuel cell' parcel containing dried pine needles mixed with wooden lollypop sticks and graphite pencils that had been splintered in a blender and then soaked in lamp oil. These were mixed with cotton wool, crushed charcoal and birch bark flakes, then doused with hairspray, wrapped in tissue paper and tied at each end so that the whole thing looked like a spliff from hell. My advice: stand well back when you light that one, and don't let it near a student on a Friday night.

The real joy in using a Kelly Kettle, however, is not related to urgency; it's about romance. It's more rewarding to search for kindling than it is to carry a tin of super-combustible firelighters, and it's more relaxing to watch the kettle gradually build to a boil. It makes every brew different and special.

Many years ago I camped in an orchard in Worcestershire. There were twigs all around the fruit trees, which made for a different Kelly experience every night. The smoke produced by these twigs smelled like vanilla, honey and almonds. I was able to savour an 'apple wood brew', a 'cherry wood brew', a 'plum delight' and a 'peach treat', all by selecting different types and blends of wood. As I sat and drank my tea, I knew that I was enjoying what A.A. Milne described as "proper tea" insomuch as "a Proper Tea is much nicer than a Very Nearly Tea, which is one you forget about afterwards".

When collecting kindling, every quest becomes a personal challenge and a special memory bonded to the location. Once, when camping, I was woken in the middle of the night by a torrential thunderstorm. I felt cold and achy, and concluded that a cup of tea was needed. I went out into the rain and, by torchlight, found some perfectly dry — and very flammable — dead ivy stems that were clinging to the underside of a fallen beech tree. They boiled the kettle quickly and

I remember gazing out through the door of my tent, with cup of tea in hand, smiling in the knowledge that I'd defied the elements. Deluge or not, Kelly was victorious.

Henry Thoreau said, "The fire is the main comfort of the camp…It is as well for cheerfulness as for warmth". This is what makes Kelly so special. She's a campfire in miniature that brings people together. Comforting and bringing cheer, she leaves us moist-lipped, warm-handed and wanting more.

See, I told you she was naughty.

May

VII

A SUITABLE AGE FOR A BEARD

*"He that hath a beard is more than a youth,
and he that hath no beard is less than a man."*
William Shakespeare

It isn't easy looking young. Only the other day I was described (by a friend who hadn't seen me in years) as being 'the love child of Peter Pan's affair with a wide-necked bottle of formaldehyde'. "Fennel, you've barely aged," he said. "What's your secret? Regular baths in the fountain of youth, or clothes-pegs tucked behind your ears?"

It's always been like this. I remember being at senior school, awaiting the results of my final exams, and having my head patted by girls from the first and second years. "Ah, isn't he cute," they said. (Such was the torment of being less than five feet tall when I left school.) At sixteen I had to present my student ID to borrow books from the library; at seventeen I was offered a child-seat by my driving instructor; and at eighteen I faced the humiliation of being refused alcohol at my local pub. And then I reached puberty. Late, I know, but just in time for my 'coming of age'. My voice deepened, my shoulders broadened; I grew twelve inches in nine weeks. I barely slept because of the pain in my bones. But all was not lost. Girls started taking an interest in me and soon I discovered how pleasurable a 'growth spurt' can be. The issue of looking younger than my years was resolved. Or was it?

At twenty-four I decided that I needed to buy a razor. The downy fluff on my chin had started getting longer, and I was at risk of being mistaken for a dandelion seed head. I got the bus into town, dreaming, as I travelled, of owning a proper shaving set: one with badger bristles and ceramic handles, where powder is mixed into foam in a bowl, and has a blade sharp enough to cut wings from a gnat in flight. But I only had a pound to my name, so I headed to the discount store and purchased a pack of five plastic razors for 99p. I returned to the bus stop and stroked my chin. The fur on my face seemed too fine to warrant using shaving foam. So I removed a razor from the packet and started scraping away at my cheeks. By the time the bus arrived, a crowd had gathered to see a red-faced rash-freak try to negotiate a lift home with only a penny in his pocket. I walked back to my parents' house with my shoulders slumped and my neck burning.

At twenty-eight, I was shaving once a week and feeling pretty confident in my newfound manliness. I took this as a sign to grow a beard. It wouldn't take long, six or seven *months* or so, and I would end up with something so impressive that girls would rush to me in a frenzied state, thinking I was a lumberjack with a hefty chopper. None of the guys would mess with me. I'd look so mean they'd think I was a part-time Hell's Angel. So I put my razor to one side and let the stubble do its thing.

Five months had passed when I caught a glimpse of my beard in the mirror. I looked closer and saw what a weather forecaster might call a 'cirrus beard'. There were little tufts of virtually transparent cloud-like fluff growing from random locations on my face. They certainly weren't whiskers and, as I stared closely at them, I concluded that I'd seen more hairs on a pork scratching. Even the mirror seemed to have a disapproving look. *(As Groucho Marx said, "Don't point that beard at me, it might go off!")* So I reached for my razor and shaved until my face was silky smooth. It took about six seconds.

The following year I found myself working for a firm that made fishing bait. I was asked to attend an event sponsored by the company. It was to be a day where celebrity anglers fish alongside and coach young anglers, thus encouraging a new generation into the sport. I was to hand out the presentations at the end of the day.

I decided to take my fifteen-year-old cousin with me, knowing that he'd enjoy himself. At the end of the event, when all the other children had gone home and the grown-ups were at the pub, my cousin and I stayed behind to pack away the display stands and load up my van. We worked quickly; keen to follow the adults to the pub. When we got there, everyone was sitting down and enjoying their beers. We walked straight up to the bar. I ordered a pint of ale. The landlady looked at me with a stern face. She folded her arms and tilted her head.

"Sorry Sonny," she said, "you've got to be eighteen to be served in here. Why don't you go and ask the organisers over there if they'll get you a cola?"

"What? You won't serve me, even a cola?" I replied.

"Don't start," she said. "I know what you youngsters are like."

My shoulders slumped and I tried not to think of the embarrassment of asking someone to buy me a cola.

"Now then," said the landlady, as she turned to face my cousin, "what'll it be?"

"Pint of Stella please luv," he replied.

This is how, at the age of twenty-nine, I ended up sitting in a pub in Worcestershire drinking cola through a straw while my fifteen-year-old cousin sat next to me, refusing to give up his beer.

Turning thirty solved the issue of me looking too young to purchase alcohol. Or so I thought. For eight years I lived in an alcohol-induced state of bliss. But then, this week, I was at the supermarket with Mrs H. We were standing at the till watching an elderly female cashier passing our items through the barcode scanner.

Blimp! Blimp! Blimp-blimp! One by one the items went through the scanner and down a conveyor to Mrs H, who bagged them while I waited to pay.

Blimp! There went my copy of The Beano.

Blimp! A packet of Werther's Originals.

Blimp! Some extra-mature cheddar.

Blimp! A pound of dry-cured bacon.

And then – silence. No frenetic *blim-blim-blimp!* of the chocolate éclairs multi-buy. Not even the embarrassing *Bliiiiiimp!* of my under-the-counter copy of *Practical Poultry*.

"Ahem!" said the cashier.

I looked up. The lady was holding my monthly-treat bottle of Hendricks Gin (the reason God gave us lips and 10,000 taste buds).

"Do you 'ave any *eye-dee*?" she asked.

"Er, no," I replied, "only my credit card."

"That wone do," she said, "I gotta see summin' with a foetoe onnit. Cos there's no way you gonna get this without any *eye-dee*."

I looked at Mrs H, whose eyes were beginning to water and whose mouth was doing its best not to break into uncontrollable laughter.

"He's thirty-eight," said my beloved, as if to break the deadlock.

The old lady looked at me, squinted and drew herself a little closer. "I suppose he is," she said, "but he can only have it if *youuuuu* buy it."

Mrs H, being the relaxed type, simply got out her purse, removed her driving licence, and showed it to the cashier.

"Oooh. Born in 1980. Ain't you done well!" said the woman. "Fancy you findin' a toy-boy half your age!"

The lady swiped the bottle of gin over the scanner. The machine made the most satisfying – and relieving – *Blimp!* I'd ever heard. Mrs H bagged the item and paid for our goods. We left the supermarket with the sheepishness of naughty schoolchildren exiting a Headmaster's office.

It's funny how growing older changes one's perception about the age of others. Young people seem to me to be growing up quicker, but look even younger; old people don't seem to age at all. (You know you're getting old when soon-to-be teenagers describe the pop duo *Right Said Fred* as being "Classic Rock Artists".) That elderly cashier must have thought I was, relatively speaking, very young.

I remember talking to my friend Bryan a couple of years ago about my frustrations with looking so young and how I felt that it was limiting my promotion prospects. Bryan, being a man in his mid-seventies and wise with life, said, "Grow a moustache. That's what I did when I was your age. It was what most of us graduates did. Helped us to look a bit older". He then gave me one of his 'knowing looks'.

"But Bryan," I replied, "I'm not graduate age. I'm thirty-six."

"Oh!" he exclaimed, taking a step back, "Then you're completely buggered!"

If looking like a 21-year-old at 36 is defined as being 'completely buggered', then what does looking like a 16-year-old at 38 mean? How is it defined? What exactly happens after one *has* been completely buggered? (Hopefully nothing more than a prickly kiss and a farewell handshake.) Not wanting to find out, I decided to do something about my baby-faced curse. As I saw it, I had three options:

1) Quit drinking alcohol. Without it, the issue would go away. *Not really an option because, as Frank Sinatra said, "I feel sorry for people who don't drink. When they wake up in the morning, that's as good as they're going to feel all day".*

2) Blue-rinse my hair, wrap myself in a tartan shawl and hop on a mobility scooter. All I'd need to do is purchase a tube of Steradent tablets with each bottle of gin and nobody would be any the wiser. *Would be okay for a while, but there's a risk I'd become fond of the scooter.*

3) Wear a beard. Not a fluffy cats-bum of a thing like I had in my twenties. But a proper beard. A stupendous beard. One that could be trimmed and groomed, and twiddled and waxed, and stroked and fondled, and aimed at anyone who challenged

my right to a tipple or two.

Being the authentic type, I couldn't entertain the idea of wearing a *fake* beard. That would be like wearing false teeth during a visit to the dentist. I'd soon be rumbled, and besides, I'd miss out on the daily updates in the mirror. Instead, I would grow a real beard and marvel at its fullness as it greeted me each morning. I'd grow one like that worn by King Edward VII. Yes. That beard had *regal splendour*; it was just the thing to wear when sipping cognac, talking to Heads of State, and getting my revenge on cashiers at supermarkets.

That was six weeks ago. I've since been growing and grooming my beard with the precision of someone training a bonsai tree. Why? Because it's not very big, that's why. I can't even say that it's small and perfectly formed. There are bald patches where I'd rather not have bald patches. Sure, I have growth on my top lip and around my chin. But on my chin? Nothing. It's as hairless as a lizard at a waxing parlour. There's got to be a solution. A chin wig perhaps? A chin comb-over? Maybe not. Maybe I'll just resort to the old 'cowardly cut-throat razor' look where I leave my neck and sideburns to grow long and only shave the main part of my face. That could work. It's traditional, rural-looking. And the best of a bad bunch – of whiskers.

So here am I, writing this as I stand before the mirror. I have a pen in one hand and a razor in the other. Should I keep writing, or should I shave off the beard?

If a beard is the mark of a gentleman, and a fine beard a debonair salute to tradition, then is a fluffy beard that clings to my face with the desperation of a Gecko in lead boots something that can be worn outdoors? What if the wind blows? What if someone thinks I've fallen asleep on a stick of candyfloss? *Stand tall Fennel. If it's what you want, and it's representative of you, then keep it.* I agree. I have, after all, developed an appetite for such things.

Beard's aren't to everyone's taste, but they're certainly to mine. And besides, looking in the mirror has reminded me of something: I'm rather fond of pork scratchings.

Stop – Unplug – Escape – Enjoy

In what ways could you
enhance your appearance
to better communicate
the real you?

June

VIII

THE CANVAS EFFECT

"The place which you have selected for your camp, though never so rough and grim, begins at once to have its attractions, and becomes a very centre of civilization to you: Home is home, be it never so homely."

Henry Thoreau

Five years ago I spent the best part of a year living in a tent beside a lake in Worcestershire. You'd think that this would qualify me as an expert camper. But my experiences proved that I'm only comfortable in the wild if I have a healthy supply of tinned beans and toilet paper. I also learned that my over-fertile mind gets the better of me at night. I blamed my tent for this. It was too attractive to wildlife. What I needed was a bigger tent, a sturdier tent. One that didn't look like a lettuce in suspenders.

I returned home and began looking for a new tent. One that could cater for all eventualities – from Arctic exploration to an African safari; something robust that could withstand the claws of a polar bear or the teeth of a bloodthirsty lion; that would remain erect in a hurricane and dry in a monsoon. And it needed to be traditional. That nylon and fibreglass rubbish reminded me too much of the famously ill-fitting bra that failed Helga Von Bustmeister during the 1932 Munich Games.

(If I remember correctly, Helga was three times World Belly Flopping Champion and part-time mascot of Save the Whale Foundation. She was reaching the finale of her 'triple-spin free-fall blubber-whack' when the bra gave way and took out three of the spectators. Helga was disqualified for entering the water a 'full' three seconds earlier than intended.)

I was compelled to purchase a better tent. A proper tent. Something built from canvas and metal. If Edmund Hillary could make it to the top of Mount Everest with a canvas tent, then I could endure the worst of the British weather in one. It was just a matter of finding the right model.

I find that doing my homework in these circumstances can prevent me from spending too much money on things I don't need. Mrs H admires me for this and knows that I can always be relied upon to make informed purchases. I am never influenced by impulse or whim. I always buy within budget and insist that my lifestyle enables me to make full use of the new purchase. *(I also have a pair of rocket-powered roller skates that run on aviation-quality bullshit.)* So I decided to do things by the book. I purchased the best guide to camping that I could find. It appeared to have been written by a retired Antarctic explorer who was now 'doing the rounds' at suburban campsites. The book contained advice ranging from 'how to survive an avalanche' to 'how to cook your morning sausage'. It was my sort of read: serious to some and completely bonkers to others. The book did, however, offer some valuable advice. It stated that the would-be camper can look forward to "spontaneous mini-breaks" where one can "roam, commune with nature and enjoy the free, wild and untameable outdoors". How wonderful. "All that's required," said the author, "is a tent, a sleeping bag, a mattress, a hurricane lamp, and cooking equipment." *(Crikey. No mention of a toothbrush, deodorant, change of undies, midge repellent and bear trap. This author was hard core.)* "And when it comes to tents," said the author, "one should purchase the lightest possible model. A four-and-a-half pound tent is the upper weight limit; any more and it will limit the range you can walk in a day." Four-and-a-half pounds? I'd be carrying more than that in tea leaves. Did this author have no style? I put the book aside and went back to my original brief, concluding that I needed a truly *magnificent* tent.

A magnificent tent is, in my book, made from canvas and has a centre pole that can double as a maypole. It has enough space inside for a 'knees up' with a troupe of Morris Dancers but packs away small enough to fit into the boot of a car. I thought that finding such a tent would be easy. But not so five years ago. There were plenty of canvas tents available in Army Surplus stores, but these seemed best suited to someone who wanted to park a minibus inside them. And there were plenty of second-hand tents for sale online. I actually bought one, travelling to North Wales to collect it, only to find it being used as a kennel for an incontinent Great Dane. (It smelled so bad that I had to stop the car on the way home and dump the tent in a dustbin.) Then I discovered a specialist tent manufacturer who still made canvas tents. They had everything from classic 'A-Frame' and bell tents to marquees large enough to house a wedding reception. But alas, all their tents were out of my price range. (I wanted something with presence, but which didn't require too much explaining to 'she who does the accounts'.)

I nearly gave up the hunt, and was on the verge of purchasing a backpacker's nylon tent, when Mrs H told me to stick to my guns and be patient. Good thing, too, because the 'Glamping' (Glamorous Camping) bug soon hit the nation in response to those seeking a bit of Enid Blyton magic in the countryside. Soon there were canvas tents being made again, in proper traditional style, at a reasonable price. Hallelujah for the barminess of the English. But I was concerned. I liked them, but my 'vanity of youth' had an issue with the image factor: that Fennel ("he of wild camping fame, creator of the MacFennel Challenge and sufferer of the Squirrel

Vindaloo") would be accused of being softer than a Glamping duvet set. That I'd be labelled as someone who preferred to eat scones with jam under canvas on a summer's day than spit roasting some roadkill over an open fire while sheltering under a tarpaulin. *(Buy it now: the Fennel Grylls sterling silver camping set with tea light scone-warmer.)*

And then I found something: the most amazing tent I'd ever seen. A cross between a tent and a tipi. Imported from Scandinavia by Nordic Outdoor (famed for their Gransforth Bruks axes and rather kinky-looking 'Double-sided Hand Strops'), the 'Tentipi' range in khaki fabric had me drooling and murmuring: "I want, I want". But again, the tents were too expensive. They were ten times more than my budget would allow. And yet I couldn't stop thinking about them. They were definitely the best, but buying one would require me to sell the car and do the washing-up for a year.

I have a hypothesis (not concluded, but definitely under study) that people who love Fine Things have a small hamster in their head (mine's called Gerald). This hamster has a habit of disconnecting the plug of all rational thought, then reaching for a set of bongos and banging out a jungle beat while singing "yeah, baby; let's do it!" This is why I found myself having a conversation with Mrs H that went as follows:

"Sweetheart, I've been thinking. Money's tight and we ought to cut back. As I can work from home and you've got a reliable car, it would make sense for me to sell mine and free up some cash."

"Sounds logical," she replied. "It would help us to clear the credit cards and maybe, if we've any money left over, we could replace the hoover that you killed by sucking up that tin of paint you spilled in the garage."

Diddy-bum-bum-bum-diddy-bum-bum-bum.

"Good plan. It's a good car, too. Should sell quickly. It's only done 200,000 miles and so long as we sell it on a cold day, no one will notice that the passenger window doesn't open and that the carpet still smells iffy from when we took Limping Pete to the chiropodist. We ought to get a few grand for it at least."

Bum-bum-bum-bum-bum-bum-diddy-diddy-diddy-diddy.

"As usual, leave it with me to sort out; just don't go spending money we haven't got. I know what you're like."

"Would I ever?"

Oooh yeah, oooh yeah. Oh baby! Let's do it! Let's do it!

The 'Tentipi Safir 7 cp' arrived via courier the next day. It was delivered to my next-door neighbour, who bought it round to my house while Mrs H was at work. Naturally, I'd also bought the essential accessories: comfort flooring, Eldfell Pro stainless steel stove and chimney, a spiral clothes dryer for the rare occasion when I might fall through ice and need to drip-dry my clothes in the tent, and a reindeer-hide rug for when

I wanted to dress for a pagan ritual. Oh, and I treated myself to one of those axes and a hand-strop spanking board as well. (Hoping that, if Mrs H was going to punish me, she'd use the latter.)

I was ready to once again be a serious camper. All I needed was eight-foot of snow, a jar of pickled herring and a bottle of schnapps.

The tent came with a label. It read: "The world's most versatile tent – for people who need maximum performance and flexibility. The Tentipi Safir is in the top 1% of the best tents worldwide". The top one per cent! *Worldwide.*

"Oooooh. Oooh baby. Gerald, you beauty! This is why I've waited five years; it's what I've dreamed about, and what I've needed. How could I have existed without it?"

The phone rang. It was Mrs H.

"I've contacted some car dealers," she said. "Your car is worth about three hundred pounds. It seems the mileage is high."

"Three hundred pounds?" I replied. "For a twelve-year-old classic motorcar? They must be joking?"

"Sorry babe, that's all I could get."

"Did you mention about the driver-side air conditioning and customised carpet?"

"No, but I don't think it would have made any difference."

Oh dear.

Bum.
Diddy.
Bum.

"Sweetheart," I said, tentatively, "I'd assumed the car would be worth ten times that. And, given the likelihood of a quick sale, I decided to spend the money now on an ultra romantic gesture. I know you urgently need a new hoover, so I've bought you a top-of-the-range rug to test it on when you do finally get one."

"What?!"

"It's a quality rug, natural hide and recently imported from Scandinavia. Obviously it came with some extras, but I'll find a use for them. I just wanted to thank you for being so utterly amazing, and lovely, and considerate. You're such an understanding wife. I love you."

"Fennel. What have you done?"

"Er…remember that trip I had to North Wales? Well, when you get home, you'll find me in the garden. Confined to the dog house."

JULY

IX

X PLUS ONE

"Once it was inconceivable for anyone to venture into wild lands without a sensible knife."

Ray Mears

It was bushcraft expert and ex-French Foreign Legionnaire Andy Sargeant who first introduced me to the term 'X Plus One'. I'd been watching him demonstrate various fire-lighting techniques at a bushcraft show and, having been impressed by his seemingly effortless ability to produce fire by friction using a bow drill, I approached him for some tuition.

Andy handed me the bow and drill and explained the principles of how they work. I got down on my knees and rapidly fumbled with the bow as I attempted to emulate the master. In my amateur hands it kept springing free and refused to build any friction. But I kept going and, while cursing and sweating, overheard Andy talking to one of the other instructors.

"Now," said Andy to his peer, "take a look at this new knife of mine. It's particularly useful for disembowelment, especially at night. Let's say I was holding a...er...'sheep's' throat and wanted to quickly dispatch the animal. The shape and length of the blade are ideal, and its black coating makes it easy to conceal when approaching the victim...er...I mean target...I mean prey."

"Yeah, yeah," replied the other instructor, clearly having heard it all before, "like none of your other knives could do that? You just wanted to buy a new knife. How many do you have now? A hundred? Two hundred? You're knife obsessed!"

"X plus One," said Andy, "X plus One..."

I stopped what I was doing, looked up at the instructors, and asked: "Not that I'm fearing for my life or anything, but would one of you explain to me what 'X plus One' means?"

Andy smiled at me and said, "The correct number of knives to own is 'X Plus One', where X equates to the number of knives you currently own".

I thought for a moment of the things I've collected over the years – hats, fountain pens, books – and realised that Andy's rule applies to them too: that no matter how many we have, or how contented we feel, there's always the temptation to have 'more'. And the latest one is always the best, until we get the next one...

"Do you seriously have two hundred knives?" I enquired.

"Knives being what they are," came the reply, "dictate that you have to be serious about them, no matter how many you have."

"Was that a political way of saying 'yes'?"

"Erm, no it wasn't. It was my way of avoiding the question."

"So you do have two hundred knives then?"

"At the last time of counting, I had 'X plus One', because there's always another one on order; always time to 'spank the Heinnie'."

"You knife tart!" said the other instructor, "When normal folk talk like that they're referring to something sexy. With you it's always about knives!"

"What could be sexier than a new knife from Heinnie Haynes?" said Andy. "Browsing their website sure beats looking at nudie girls jiggling their bits about; at least online."

"It's all smut to you," said the instructor, "all X-rated."

"I do like black," said Andy. "Or as I call it: 'blacktical'. But it doesn't have to be in leather. Kydex or bolatron is fine. It's all about *the edge* anyway, about how sharp I can get it."

The other instructor turned to me, winked, and then whispered: "He's going to talk about Obsidian blades now".

"Sure, you can get 'razor' sharp blades that can cut hairs hanging under their own weight," continued Andy, "but what about 'surgically' sharp blades that can pass between the cells of your skin without tearing them? You can cut yourself and not even feel it. It takes some sharpening and stropping but you can get there in the end. But the ultimate? An Obsidian blade: a form of glass that when knapped was once the most prized tip for a spear. Its point is so sharp that it's measured at a *molecular* level. It can slice a single cell in half! No wonder it's used by eye surgeons."

"But it's hardly much good at carving wood or striking a fire steel, now is it?" said the other instructor. "Isn't it time you helped our apprentice? He hasn't got a smoulder yet."

"Or a knife!" exclaimed Andy. "That kid needs to tool up!"

I stopped what I was doing and wondered where I'd put all the penknives I'd accumulated over the years. Somewhere at the back of my garden shed was at least one Swiss Army knife. And I could remember owning a pruning knife and a letter-opener that could just about cut through paper.

When I finally got home I started rummaging about in the shed, ultimately finding seven knives in various states of neglect. There was nothing 'blacktical', 'sheathed' or 'stropped', just a bunch of rusty and unloved knives that had mostly been used as makeshift screwdrivers. Surgical sharpness? These were blunter than a Glaswegian debt collector. It wouldn't do. It just wouldn't do at all.

I felt hopelessly unequipped. I needed a new knife. Project 'X Plus One' had commenced. Time to research what was available. Nothing expensive, just something cheap that I could use to practise my sheep gutting skills. Having learned my lesson with the tent, I would only allow myself to purchase a good one when I'd reached the level of swordsmanship displayed by a Samurai warrior.

Hmm. That sort of approach has never worked for me before? But I'll go with it, for now…

Initial reading (of Mors Kochanski's *Bushcraft*) taught me that for bushcraft use, a sheath knife is best. (Unlike folding knives, they don't buckle at the wrong moment and chop the user's fingers.) I also learned that the blade should be 3mm thick, made of carbon steel, and be approximately the length of the width of your palm. It should curve along its entire length to a point which should be in the middle of the knife's centre line when looked at from the side. The blade and handle should be made from one continuous piece of metal, known as 'full tang', which gives it strength right down to the 'pommel' at the end (which can be struck so to drive the knife tip into something) and the 'scales' of the knife's handle should be made from a material that can be carved or sanded to fit the profile of one's hand. Finally, the profile of the knife should be that the top of the blade and handle follow the same smooth line, and the top of the blade should be wide and square to make it strong enough to be hit with a wooden club when felling a small tree. (The ultimate test of a knife's strength is that it can be hit from above and driven 4cm into wood without breaking.)

My old penknives clearly didn't meet the required specification, no matter how much I sharpened them.

I visited the Heinnie Haynes website and was immediately overwhelmed by choice. There were hundreds of beautiful knives, many of them bespoke pieces. But try as I might to rationalise my decision-making, I found myself drawn to the knives' aesthetics as much as their functionality. All the black military-looking knives? Urrgh. They weren't for me. Neither were the fluorescent 'zombie killer' things. And handles made of plastic? They wouldn't do either. I dismissed them in the same way as I retract from disposable lighters and Biro pens. No. They wouldn't do at all. I needed something traditional-looking and handcrafted that would bring me joy and last me all my years.

After two hours of searching, I came to the conclusion that I ought to go straight to TV's top bushcraft expert for advice. No, not Bear Grylls (he sounds like a Canadian franchise of the Angus Steakhouse) but Ray Mears whose programs I'd watched for years.

Luckily for me, the Ray Mears website sold just the knife I was after: the *Ray Mears Bushcraft Knife*. Hand-made by the revered British knife maker Stephen Wade Cox, it ticked all the requirements laid out by Mors Kochanski and looked sexier than Kelly Brook playing Zero-G volleyball. It was 'added to basket' before I'd spotted the hefty price tag, but if it was going to be used for whittling wood then it might as well whittle away my savings.

The knife arrived two days later in a stunning presentation box accompanied by three Japanese sharpening stones and a bottle of Camellia oil (which, at £17.00 a bottle, smelled worryingly like WD40). I removed the knife from its tanned leather sheath and held it in my hand, feeling the perfect balance and noticeable weight.

Here was a knife designed by an expert and made by someone of equal skill and understanding. The English oak handle pleased me immensely, being most appropriate for me as an English countryman. And yes, it was something I was proud to own. With luck, I would use it on many an adventure and in doing so I'd be respecting its makers – something that seemed far more important than the exchange of money.

My new bushcraft knife would be the grandaddy of all the knives *that followed*. For even as I was marvelling at the Ray Mears knife, I was thinking about the next ones: a bigger one and a smaller one. Maybe even a folding 'every day carry' knife?

But wait a minute? That's three extra knives. The saying isn't true at all. There's no 'X Plus One', at least when it comes to Fine Things. Here 'X Plus Three' applies. And we're more the merrier for it!

Just don't tell the missus…

Stop – Unplug – Escape – Enjoy

What 'one more thing' have you always dreamed of owning?

July

X

BOTANICALLY BREWED

Mrs H has decided that she and Little Lady will be spending a few days at her parents. Apparently this has something to do with me spending all our savings on a tent and a knife. Next month's mortgage payment is at risk, and I'm challenged to find the money.

How ridiculous. I mean: if we're going to lose the house, then we'll *need* a tent. A big tent. One that can accommodate a family of three, plus forty-seven flat caps, three-hundred countryside books, seventeen fountain pens, a rucksack, some beard scissors, and a shiny new bushcraft knife. And yet I'm the one being blamed for our predicament. I was thinking ahead, that's all. But Mrs H has got the hump. She and our daughter are to have a mini holiday, being pampered by nana and grandad, and I am to remain at home. Working. I have strict orders to finish writing the books that might earn us a few pennies. I'm not to answer the door, or the telephone, or open any letters. I am to remain focused. Disciplined. *Productive.*

Unfortunately for Fennel Enterprises, I'm the biggest procrastinator in the whole kingdom of Faffdom.

(Faffing being the skilful art of putting things off, or being easily distracted by things of low importance.) Not having Mrs H around or, more accurately, not feeling her boot against my backside, is going to challenge my ability to do any writing at all. I will, after all, have the house to myself.

Like most adult males, I have only one priority when my other half is out of the house. I lock the front door, close the curtains, and raid the fridge. Also, shameful addict that I am, I plunder my secret stashes...for a much needed drink.

When Mrs H and I started dating, she asked me if I was financially stable and whether I had any addictions. (She's very practical like that.) Of course, I lied on both counts. My debt rivalled that of the Greek Government and I was suffering from a twenty-six year drink problem. Not to alcohol, but to Fentimans' ginger beer. So when she eventually learned of my addiction, she overcame her horror by rationing my weekly supplies of Fentimans. I was allowed to drink no more than five bottles per day, and would only be allowed to chill one bottle at a time in the fridge. I got round the restrictions at first, smuggling extra bottles into the house using a fake briefcase. (As a gardener, I had no need for a briefcase, but I got away with it on the pretence that I was 'career minded'.) I found some pretty good hiding places for the bottles, too. One was tucked behind the sink pedestal in the bathroom, a bottle was placed into each of my wellington boots, and one stashed in the laundry basket. (Nobody ever goes in there, right?) But then I tried to be too clever. I stored eight bottles in the toilet cistern which, when Mrs H flushed the loo, released a mere dribble and caused her to investigate. And then the ultimate disgrace: Mrs H learned of my secret Post Office Box – the one used for my forbidden 'wholesale' deliveries. Royal Mail had sent me a letter, complaining that I was abusing the facility. Mrs H found the letter on my desk, went to see them, and came back with fourteen crates of bottles. Oh how I cried when she poured them down the sink. But that was years ago. I've since mastered the art of concealment, using the water tank in the attic as my ultimate hiding place. Mrs H never ventures into the loft, and the water in the tank is always cold, so I have a reliable supply of cool ginger beer for the foreseeable future. (Just don't tell her, okay?)

I have no desire to offend my beloved, which is why I've waited for her to leave the house before I quench my thirst. So, as I 'anticipate the moment', I can confirm that the door is locked, the curtains are closed, the fridge is raided of the one legitimate bottle of ginger beer, I've been into the attic and removed thirty-six bottles of Fentimans from the water tank, and I've even collected my favourite Norman Wisdom box set from the DVD shelf. I'm now sitting on the sofa, readying myself for eighteen hours of viewing and drinking pleasure.

I have no shame in what I'm about to do.

Doing all this is, technically, preparation for writing the book. And it's allowing me to write these words. So I don't feel guilty. I'm just putting myself in the right frame of mind, that's all. Coleridge took opium to create his Kubla Khan, and I shall resort to Fentimans to expand my mind. Because, and I warn you in advance, it has far-reaching effects on my creativity.

Fentimans, if you don't know, is a non-alcoholic *botanically brewed* beer. This traditional technique, used by company founder Thomas Fentiman in 1905, involves milling ginger roots into a paste, fermenting them in a mixture of water, glucose syrup and yeast, and then tumbling them into heated copper pans. Pear juice is then added, and the broth simmered. It's flavoured with lemon juice, capsicum, more ginger, and a herbal infusion of speedwell, juniper and yarrow. The non-alcoholic beer is then transferred into wooden vats, where it's left to ferment. The result is a ginger beer with enough kick to floor a giraffe.

Sounds nice, doesn't it? That's why it's my first and last choice of ginger beer. Although, much to my embarrassment, Mrs H has taken to using the term 'botanically brewed' to describe my wackiest behaviour, which usually comes as a result of a session on the Fentimans. So here goes...

I've decided to begin the session with a bottle of Fentimans and Norman Wisdom's *The Early Bird*. Drinking from the bottle means I'll get more of a bubbly hit, and when the bottle's empty I can blow across its lip to create the night's musical entertainment. First, though, Mr Wisdom has to make a cup of tea before he's properly awake. It's one of the best comedy movie scenes of all time. And it involves tea, which makes it my sort of film. I'll put down this pen and come back to you once the film has ended.

Okay, I lied. I am halfway through *The Early Bird*. Norman is on the golf course, dressed as a vicar and swinging from a tree, and I've found myself with an empty bottle of Fentimans in my hand. I forgot to put any more in the fridge, so I'm going to have to drink a coolish one. This time, however, I'm going to create a *Fentimans' Slurper*. It involves pouring two bottles into a breakfast bowl and, cupping the bowl in two hands, raising it to one's lips and sucking the top layer of beer from the bowl. If I get it right, the beer should vaporise as it enters my mouth and the ginger will taste that much stronger. Get it wrong and I'll choke a mouthful of ginger froth. It's worth the risk though, for a stronger 'hit'.

Update. I'm four *Slurpers* down and three films in. Norman has saved his milkround in *The Early Bird*; he's hobbled on stilts to become a policeman in *On the Beat*, and foiled a robbery in *Trouble in Store*. I've laughed and slurped, slurped and laughed. I've slurped and slurped, and slurpety-slurp-slurped and laughed. I've slippy-slurp-slurped, and burped, and belched, and laughed, and felt like a man in charge of his domain. Ah, yes. I've ruled my ginger kingdom. Master of beers. Brewer of gawd-knows-what in the morning. And then, when the film's gone quiet, I've *sloooooooeeeeeerrrrrrrppppped* to the bottom of the bowl and given myself a sugar rush that's had my eyes blinking and my teeth chattering. Eeeee. T'was great Mr Grimsdale! Though it seems to be dizzying my thoughts. Not to worry. I'll keep going. Twenty-seven bottles left.

Ooh. There's that bottle that I finished earlier. It's empty. C'mon, play a tune with me! Let's do *The Great Escape*! *Choo-choo, choo-choo-de-choo-choo.* You're not blowing. I meant out loud. Try harder. *Choo-choo, choo-choo-de-choo-choo.* Is it just me? C'mon! It doesn't have to be in tune. Those airmen won't escape by themselves. They need our help. *Choo-choo-* oh, forget it. Time for some shots.

Fennel's Fentimans' Slammer Shot was invented at 2.47am on February 6th 1993 when, after getting bored with Tequila slammers and shots, I decided to make a combined ginger beer equivalent of these popular party drinks. It involved a tumbler glass, sugar, a root of fresh ginger (finely sliced lengthways and soaked in a bowl of ginger beer for half an hour), and a bottle of Fentimans for each pocket about your person. The process of drinking it is a merged version of the Tequila Slammer and Tequila Shot:

1) Half fill a tumbler with ginger beer

2) Briefly lick the skin between your thumb and forefinger of your left hand, then sprinkle a pinch of sugar onto the area

3) With the same hand, remove a slice of ginger from the bowl, holding it between your thumb and forefinger

4) Place the palm of your right hand over the tumbler, and hold the glass tightly with your fingertips (you want to ensure an air seal over the glass)

5) In quick succession, lick the sugar from your hand, lift and slam down the glass onto a sturdy surface so that the beer fizzes up, drink it quickly and then suck as hard as you can on the marinated ginger slice, swallowing the entire 'hit' of sugary gingeriness. Then await the biggest burp of your life.

Repeat the process until all slices of ginger are used up, then (optionally for the serious ginger beer drinker), do a *Fentimans' Slurper* with the infusion that remains in the bowl. It's enough to make your head spin 360 degrees and your arms flap while you jump uncontrollably around the room.

Okay. The *Slammer Shot* is ready. My hand is sugary and I'm holding some limp ginger up to my mouth. Ready, ready? Okay, go! Sugar-slam-slurp-gulp-suck-gulp.

Oooh-eee-jeee-a-chu-kuu-ju ju ja-ja

That hit the flot. Ooh. My thongue's gon thumb. Thime for another.

Grr-awooo-cha-cha-cha.

Blummin' heck. My eyelids have started to twitch. Still, I can see my way to drinking one more.

Hoo-ja-hoo-ja-ay-aya-ay-ay-ya!

Okay. Bleathe. Clalmly. Try to flocus your eyes Flennel. And drink annother.

Hurrr-ubba-ubba-dubba-dubba-doo-doo-grrr-eeee-na-na-na.

Oh, er, no. I'm not quite done.

Hooooo-ha. Hoooooo-ha. Krrrr-jibba-jibba. Icky-icky-na-da-na-da.

Ooooh. Oh yeah. Oh baby. That did it. All I have to do now is await the burp. Oh. Here it comes.

Baaaarrrruuuuuuuoooooooouuuup!

Cor. Heck. That was gingery. Apologies. I just had to 'clear what was within'.

Five *Slammer Shots* down and a few more to go, I've decided the mood is right to watch *Stitch in Time*. This is classic Norman Wisdom. It deserves the full treatment. I wish to engage with the film. I'll nip upstairs and change into my pyjamas and dressing gown.

Right, I'm back. And I've picked up my flat cap and put it on my head. I also realised that the classic sketch in this film involves Norman getting wrapped in head-to-toe bandages. So I'll start the film and begin wrapping myself in toilet paper bandaging. That ought to do it. Oh yes. This is the film. A hold-up at the butchers followed by some serious on-ambulance adventuring. Bring it on!

Which reminds me. Did I tell you that I once got stuck on top of an ambulance? No? Oh. Probably because it never happened. Still, there's always time. But as a start, I'll climb up on the back of the sofa and flail my arms about a bit.

Cor. That was such fun. Not done that in weeks. And while watching such a good film, too. But it's time for the finale. My favourite-ever Norman Wisdom film: *The Square Peg*, just in time to slurp what's left in the *Slurper* bowl. I'm going to keep the bandages on but grab a shovel and my work boots from the garden shed. I'll then drop behind the sofa and reappear just in time for the parade scene where Norman's shouting alternative commands to the soldiers.

Right. We're here. Behind the sofa. The film's playing and I'm going to do my best 'Hannibal Lecter slurp' to finish off the ginger beer in the bowl. Here goes:

A-slith-slith-lith-lith-lit-lit-lit-lit-lith-uuuuurp.

There they are! In the parade ground. Are you ready to jump up? Okay, after me. One, two, three!

"Slow! Slow! Quick! Quick! Slow!"

"Slow, no!"

"Left, left, right, rig-"

*Oh. Dear. Oh No. Oh *****!*

Erm. How can I put this? We have a situation. A visitor. Mrs H. She's back. The door's opening. She's in the room. Hide, quick, duck behind the sofa!

"Hello?"

Ignore her. She'll go away.

"Erm, Fennel. I know you're behind the sofa. Why don't you come out and explain to me why the living room floor's covered in bottles, DVD boxes and empty toilet rolls?"

We'll have to do it. There's no 'Great Escape'. I'll show my face.

"Ah. As I thought: Norman Wisdom. I can tell by the way you're wearing your hat. Well, come on! Stand up when I'm talking to you!"

She's left us no choice. We'll have to comply.

"Gawd! What…on…earth…are…you…wearing? And what's that in your hand? A shovel? Oh, no. You haven't? You have, haven't you? You've been on the Fentimans again!"

Hmm. Seems I have some explaining to do.

Botanically brewed, or right royally screwed? You decide…

AUGUST

XI

THE PEN THAT FLEW THE ATLANTIC, TWICE!

Ten years, it seems, is the official threshold for when a man's wife takes ownership of the family jewels. Until then, 'real men' convince themselves that they're in charge by pretending to wear the trousers of authority. With their skewed version of reality, they think themselves the alpha males who keep their good ladies on a short leash. The reality of relationships, I'm sure you'll agree, is rather different. Women have been in charge from the beginning.

My marriage to Mrs H is a deeply loving one. I'd be lost without her, especially now that she and I have a daughter. My wonderful wife is the single most beautiful, caring and supportive person I've met. Always putting others before herself, she has more patience and tolerance than I deserve. I am, I admit, a difficult person to live with. I'm a classic eccentric, living at the extremes of high mania and low mood. There's no middle ground, only madness and sadness. Neurotically insecure at times, yet invincibly confident when I'm flying high, I battle my way through self-doubt by sustaining my high creative output. If I rest, or stop to think for too long, I'll very quickly need Mrs H to pick me up again. How ironic for an author who encourages others to 'Stop – Unplug – Escape – Enjoy'. It's what I always need, but sometimes fear. The bright light of brilliance keeps the darkness away, but it can be so very exhausting. Hence why I'm so impulsive with my purchases. I'm attempting to cheer myself up, even though I know the pleasure will be short-lived. Always seeking to understand myself and improve my image, I hold faith that so long as I'm true to myself, and invest in my future, life will always get better. Mrs H sees things differently. Well she would, wouldn't she? She does the accounts.

Mrs H has forgiven me for my recent 'investments', even for my private ginger beer moment, and is home for good. But on one condition: that I destroy my credit cards.

I am distraught, knowing that my spending vice is over. Gerald's bongos have stopped and, as I approach mid-life, I feel like the old boot that lands on Mayfair after an eight-hour game of Monopoly.

Being the mature and responsible adult that Mrs H hopes for me to be, and knowing that there is only one way through a crisis such as this, I agreed to her terms and handed over my credit cards. As I watched her cut through them with scissors and throw them on the fire, I felt the crushing weight of 'freedom' press down upon me. But, whilst Mrs H thought that my spending spree was over, I knew that my biggest-ever purchase would soon arrive in the post. One last item, just for old time's sake.

Bum. Diddy. Bum.

There's always been an ultra-special item at the top of my wish list, a luxury item that – should I ever win the Lottery – I would purchase as my 'one thing to rule them all'. The item has changed over the years, but for the past five years it's been constant, proving that it's my ultimate – and totally unaffordable – desire: a Sailor Susutake Smoked Bamboo fountain pen.

Handmade by the now-retired master pen maker Nobuyoshi Nagahara, each pen is crafted from unique and incredibly rare bamboo thatching that was salvaged from ancient Japanese houses prior to their redevelopment. The colouration of each pen is unique, developed from over 150 years of being gradually infused by the smoke from the open fire of the house. They have a pure gold nib, with the finest build quality in the world, are inscribed with the maker's name and are presented in a kimono-style silk sleeve. Just what the avid fountain pen collector would want and exactly what an angler known for using bamboo fishing rods would desire. The price for such a beautiful item – so I am told – is 'irrelevant'. But if pushed, I'll confess "it's definitely not over two-thousand pounds – darling".

How sweet the lie can be when the reward is so great.

One has to react to one's situation, doesn't one? With no credit cards left, I'd have to be creative with my source of funds. For example: 'just suppose' Mrs H had put her life's savings into our joint account so that she could purchase a new kitchen. What if someone borrowed that money with the full intention of putting it back 'if' its owner discovered it missing? What if the funds were invested wisely, say, in something rare and collectable? Like, and it's totally off the top of my head, a bamboo fountain pen. Perhaps one of those Susutake thingies? You know, the ones with a gold plate on their side bearing the maker's signature, which are smoked to a chestnut brown and have a unique bamboo node that provides the perfect balance and grip? Just like the one available from a dealer in America, that's on 'Buy it Now' for a mere £2,300? It's rarer than unicorn ivory, available for a limited time only, and there for the taking. What do you think? Now or never?

Bum-bum. Di-diddy-bum.

If we were to temporarily 'reallocate' funds to secure the pen, and ignored the warning in the maker's name (No buy o shi-), would it have a positive or negative affect on my marriage? Hmm. Want to take bets as to whether my other half would find out?

"Live dangerously", that's my motto. (Actually, it's more like 'Shop – Unplug – Escape – And Face the Consequences'.) "In for a penny, in for two-thousand pounds."

The golden mist of opportunity rarely presents itself so vividly, so I decided to buy the pen. Mrs H was too enraged by the spending on my credit cards to worry about her own savings. She wouldn't even look at our joint account, now would she?

Wives are unpredictable creatures. How was I to know that Mrs H would speak to a kitchen company on the very day that I transferred the funds from our joint account? Or that she'd view our online balance on the same day I received the shipment confirmation from the pen retailer in America. How would I know that her scream would threaten an avalanche in the Himalayas, and that she'd immediately be on the phone to our bank informing them of an "Effing Arstard" fraudulent transaction?

Oops.

Have you ever had a silent conversation with someone when they're on the telephone to someone else? It looks like a cross between sign language and charades, with the facial expressions of someone who's just realised – mid-brush – that they've mistakenly put haemorrhoid cream on their toothbrush. There's an eagerness to know, and an equal desire not to know, what's being said. And so it was that I attempted to tell my good lady that our account had "not – been – hacked!" and that it was just a "simple misunderstanding", Mrs H made her apologies to the bank, put down the phone and, with two arms reaching out for my throat, said,

"Fennel, WHAT – HAVE – YOU – DONE?"

"Darling," I said, trying not to sound too excited, "I've bought this pen. Well, not just any pen, it's a-"

"Don't tell me," she snorted, "a pen that's been on your wish list for ages and which you just 'had' to have?"

"Well," I replied, "now you mention it…"

"But Fennel," said Mrs H, lowering her hands, "That's hardly a problem in the bigger scheme of things. Our joint account has been hacked by someone in America. There's more than two thousand pounds missing. I'm hardly going to worry about you spending a few quid on a pen. Now, PLEASE, help me to find out what's happened."

"Er, Sweetheart," I muttered, reaching out my hands to hers, "what if the pen was really, *really*, special? What if it was more important than the kitchen, or any amount of fretting caused by silly things like credit cards? What if it would make me, no, *us,* really happy?"

"Could it boil a saucepan of water? Or roast a chicken? Or wash clothes? Or clean plates without covering them in stale grease?"

"No?"

"Then how could it possibly make me happy?"

"Oh," I sighed, "what if it were made from a super-rare smoky antique grass that looks like wood but isn't wood and which is too good to write with but which would look amazingly good on my writing desk?"

"I see your point," she said, crossing her arms.

"I'll stretch to a fiver. But if you've spent more than that, then the only thing to stretch will be your ear as I bend it around your head. You're hopelessly in debt, and I've got to pay for the new kitchen. So how about you stop bothering me? The bank and I need to figure out where our money's gone."

It was then that I walked, head held low, into my study and picked up the order confirmation for the pen. I returned to the living room and handed the slip to Mrs H. She held it up to her eyes, read it, and then – well, I can't fully remember. I saw stars, then everything went black, and the world stopped spinning.

That's how I came to own one of the rarest pens in the world, albeit for a day, without ever having set eyes on it. The pen successfully flew the Atlantic, but was returned, unopened, by Mrs H. The reason for the return was stated as: "Idiot". We were given a refund, but we haven't yet got our new kitchen. Well, I couldn't let Mrs H be so reckless with her money, now could I?

Stop – Unplug – Escape – Enjoy

What will you do for the loved one
who has most supported you?

SEPTEMBER

XII

THE GUV'NOR

There comes a time between friends when certain guards can be lowered and no action, however extreme, can prove shocking. I've shared things with you, so we might as well get closer. With this in mind, I'd like you to calm your thoughts, put your hands on your thighs and picture an up-close image of...my buttocks. To be precise, I'd like you to imagine them moist with perspiration and wiggling purposefully from side-to-side. Slowly.

Can you see them? They're there. Almost within reach. Do you need a little more time to focus? Do I need to clench my cheeks, or are you okay with the image you've got? Would it help if I said, "Jig-jig, jig-jig?"

C'mon, you're not trying hard enough. *Look*. Surely they're the best and most pert things you've seen all day? No? Pity. I'd hoped that they might have caught your attention.

What if you pictured the same warm, gyrating buttocks wrapped in black Lycra that clings to all the curves, crevices and dimples, but and sags and flaps in all the places I wish it wouldn't. It's not a pretty sight, is it?

Now that I've subjected you to the worst image imaginable, and have probably caused you permanent psychological harm, I am poised to make a firm promise to you: that the images you just pictured will never, ever, come true. Apologies if this saddens you, but it ain't gonna happen. There's no more chance of me wearing skin-tight Lycra than there is of seeing a swan paddling a canoe. And yet, as a cyclist, I'm supposed to wear that sort of thing. But it's not me. I'm not of that ilk. I'll never shave and oil my legs, or apply for membership of the *Squeaky Cheeks Cycling Club*.

Why am I getting so emotional about stretchy fabric and cycling? Firstly, Lycra on a middle-aged man is wrong. Plain and simple. It might be okay on muscular athletes, but on the rest of us it looks as flattering as seeing a man in shorts cutting his toenails in public. (Very pleased to meet you; although I appear to have already met the family gherkins.) Secondly, I've developed a strange perception of cycling after visiting my local bicycle shop this morning. Let me explain.

I've been worried for some time that my midriff is beginning to show signs of me drinking too much ginger beer. All that sugary-fat wobbling about on my waistline. It's not good. To make matters worse, my legs have become spindly from spending too many months sitting at my writing desk. Knowing that I'm a little too close to forty for comfort, I decided to get fit. And as I'm averse to treadmills, I opted, with Mrs H's blessing, to buy a bicycle. I'd managed to save a hundred pounds since the pen fiasco, so would spend all of it on the bicycle if I had to. With no desire for compromise, I expected to get a *really* good bike.

The last time I got a new bicycle was when Raleigh Choppers were the 'grooviest' things on the street and a BMX could help you fly across the sky. In fact, the more I thought about it, I'd never *purchased* a bike. They'd always been a gift from Santa. So I didn't really know where to get one or how much they cost. Fortunately for me, a local newspaper advert revealed that there was a bicycle shop in town. I'd pay it a visit.

The Free Wheel's Bicycle Shop was named apparently for reasons other than the price of its bicycles. As I stood outside, gazing through the window, I could only focus on the price tags and signage: "Sale Now On", "Amazing Bargains", "Prices From as Little as £800".

Eight hundred pounds? For a bike, in the sale? For that money, I'd want something that came with an engine.

I reasoned that the bikes in the window were highly specialised and that there'd be something inside that would suit my budget. I was wrong. Once inside I found that the bikes in the window really were in the sale. The first non-sale bike I saw was labelled "£1,500", the second – a garish fluorescent yellow thing – was "£1,800"; and the third was an astonishing "£2,400".

Bicycles had either become super-desirable or some marketing man was having a laugh. An assistant approached me.

"Can I help you?" he said.

"Not at these prices," I replied. "Is it usual for bikes to cost the same as a second-hand car?"

"Oh yes. But don't be put off; we're the cheapest around. We don't stock rubbish though. These are all good brands that will last a few years."

"A *few* years? Don't you mean a lifetime?"

"Not these days, mate; not with carbon."

"What do you mean?"

"Micro fissures."

"Micro what?"

"These bikes have carbon frames and wheels, and they're subjected to lots of juddering and jarring from the road. Over time the knocks can cause microscopic fractures in the carbon fibres which build and build and then, 'BANG!' the whole bike can explode."

"*Explode?*"

"Yeah, actually, the guy over there had it happen to him last weekend." The assistant called to his colleague behind the counter. "Joe, what speed were you doing when your bike blew up?" "About fifty miles per hour," came the reply. Joe then lifted his T-Shirt to reveal a graze that spanned the width and length of his back. "Other than these cuts," said Joe, "all I was left with were two tyres, a chain, and the brake cables.

Rest of the bike disintegrated."

"That never happened with tubular steel bikes," I stated.

"Aw, man, they went out years ago," replied the assistant. "When did you last buy a bike?"

"Sometime in the early '80s."

"Jeez, that was before I was born. You don't want something from back then. Those bikes were too heavy. You want something special."

"Okay, I replied. "Impress me."

The assistant pointed to a matt-black road bike with tyres the thickness of spaghetti and a saddle sharp enough to skin a deer. "It's a Shiv Di2 with crosswind-optimized airfoils, OSBB, control tower fit, fuselage integrated hydration system, ceramic bearings, TT chains, nylon balls, 2x10, reversible 12.5 mil, body geometry, blackbelt protector, and full carbon clinchers. Yours for five grand."

"Five *thousand* pounds? It looks like it could blow away at any minute."

"Whatdya expect? It's top of the range."

"Sorry. It's a little too slick for me. And I have no idea what a carbon clincher or blackbelt protector is. In fact, I have no idea what any of it is. I like things to be classically styled, handmade in Britain, and made of steel. Is it any of these?"

"No."

"Listen. I'm after something with mudguards, a

pannier, a gear lever that can be operated by my thumb, possibly a wicker basket and definitely a saddle what won't perform a Jewish operation on my manhood."

"Oh. You want a *traditional* bike. We don't stock those."

"Suppose you did, and I wanted the traditional equivalent of that helium-filled scythe over there, what should I be looking for?"

"You'd want a Pashley. They're the best. Most stylish they are, and they have the best heritage. Pashley is the oldest bike manufacturer in Britain. Their bikes would be perfect for you. Go and look them up."

I thanked the assistant, winced at the guy behind the counter, and then left the shop with the sole intention of finding out more about this manufacturer of traditional bikes. Two hours later I was standing in *The Traditional Cycle Shop* in Stratford-upon-Avon having a very different conversation.

"Would Sir be cycling in the town or country?" said the manager.

"Mostly country, though some urban cycling is inevitable." I replied.

"And would Sir favour an upright or crouched cycling position?"

"Upright when taking it easy, with the option of leaning forward when pushing for more speed."

"And what would be Sir's budget for such a bike?"

Reflecting upon recent events, and wishing to avoid

further embarrassment, I replied "About eight hundred pounds".

"Then please come with me. We have just the thing."

The man walked me to an area of the shop containing a line of the most beautifully styled bicycles. Some were Edwardian in fashion, with black frames and tanned leather seats; others looked continental with pastel-coloured finishes, willow baskets and leather pannier bags. These were the type of bikes I sought. The sort one could ride while wearing a tweed suit and a very large grin.

"Sir looks like the kind of gentleman who appreciates quality, who seeks something with refined style. But you're a young chap. You don't want something that creaks and rattles like a vintage pram; you need something with street-cred that looks the business. Am I right?"

"Absolutely."

"Then have a look at this."

The man pointed to a bike that made my mouth pucker like a chimpanzee requesting a large gobstopper.

"Pashley Guv'nor," said the manager. "Named after the owner of the firm and styled on a 1930s Path Racer bicycle. It's got a Reynolds 531 tubular steel frame, North Road handlebars with leather grips; Sylvan Stream rat-trap pedals, a Brooks 'antique brown' leather saddle, Sturmey Archer three-speed gears, Westwood rims, relaxed forks and cream Schwalbe tyres.

And it's made by hand here in Stratford. Which means, when you add them all together, that..."

"It's pure sex. I've got to have it. Monthly payments okay?"

The man smiled, knowing that the bike had sold itself. I didn't need to sit on it, or ride it, or haggle over the price. It was a dream bike. No carbon gizmos, no risk of explosion, just the best looking and most stylish bike I could have wished for. It epitomised one of my favourite literary quotes, that of H.G. Wells in *The History of Mr Polly*: "He did not ride at the even pace sensible people use, who have marked out a journey from one place to another...He rode at variable speeds... And sometimes he was so unreasonably happy he had to whistle and sing." Which is how, after a day of mixed emotions, I have come to own one of the best Fine Things I've ever seen: a Pashley Guv'nor bicycle.

All I need to do now is ride it.

September

XIII

THE SILENT ROAR

My friend, how do you fancy getting your leg over in public? Okay, maybe not in public. Somewhere more intimate. In fact, just with me. We could go down some quiet country lane, at a discrete time of day, and get all 'pumped'. And when we're done, we could go to a pub and tell the locals how far we've gone together. Sure, they'd splutter into their beers as we rubbed our sore bottoms and aching knees while talking of our gay adventure; but it's all part of the joy of being ...a rural cyclist.

Innuendo aside, I hope you'll join me for a ride some time? I am, after all, getting rather into this cycling malarkey. I've only done a few trips, mostly to the pub and back, but the pleasure of riding my new Guv'nor is delightful. It's comfy and has head-turning good looks. It's young but retro-styled. It doesn't go too fast, and it hates all that dirty off-road stuff. This so-called racing bike (whose three-speed Sturmey Archer gears would struggle to outpace a treacle-powered milk float) supports my belief that one's recreations should be leisurely. It forces me to take-in my scenery and allow the clicking of its gears to be the metronome for my thoughts. Happy times, on two wheels.

There is perhaps one style of bicycle that I crave to ride even more than the Guv'nor. It's the ultimate two-wheeled thrill-maker: a direct drive, 53-inch and 18-inch wheeled, solid rubber tyred, moustache handle-barred, thoroughly beautiful and totally terrifying Victorian velocipede known as the Penny Farthing. Built to last and never requiring a puncture repair kit, this classic bicycle is an iconic representation of English eccentricity – as deliciously daft as putting milk in tea or mint sauce on lamb, skipping round a maypole, rolling cheeses down a hill, or whacking conkers suspended on a string. It's as British as roast beef and, like the bovine Sunday lunch, makes me want to lick my lips and shout "Gravy!"

Cycling is not something I would usually holler about. I much prefer an intimate ride at the dawn of day when road traffic is lowest and the loudest thing one hears is 'the silent roar' of the breeze in one's ears. But I've recently experienced the complete opposite: a flamboyant display of cycling excess that makes me want to shout proudly from the treetops that I am a 'free-wheeler'. There. I've said it. And it's all John Summers' fault.

Hearing of my joy with the Guv'nor, John secured tickets for him and me to participate in the Tweed Run.

The Tweed Run, if you don't know, is an annual event where 500 traditionally attired ladies and gentlemen ride vintage bikes through the heart of London. They stop for tea and cake, then a picnic in the park, before bringing traffic to a halt with their exuberant displays of cap-waving, flying helmet-wearing, umbrella wielding 'chap-like' madness. There's every type of classic bike, including the much-admired Penny Farthings, many of which are fitted with bells and trumpet-style car hooters. *Ding-a-ling! Hooooo-haaa!* And 'by Jove spiffings' it's excellent fun, too.

John and I attended, experiencing the most fabulously bonkers demonstration of 'strength in numbers'. There were many thousands of people applauding us as we cycled past. So much laughter. So many classic bikes. So much tweed. Such enormous moustaches, and such eccentric behaviour. What a great 'revolution' it was.

I have two vivid memories of The Tweed Run: the first was John and I cycling side-by-side over London Bridge with three Penny Farthings towering above us. The second was attempting to talk to John as we cycled along together. You'd have thought that this would have been easy, but the only reply I could get from him was "Pear", which he repeated over and over again. Towards the end of the ride, he was slurring a hypnotic "Peeeeeaaaaaarrrrrr   " with his face sinking like a love-lost puppy.

At the end of the event, as John and I sat and drank

our glasses of complimentary champagne, I asked John to name the most memorable thing he'd seen. "Was it St. Paul's Cathedral, or Buckingham Palace, or the Houses of Parliament?" John looked puzzled for a moment, then slowly drooled, "Peeeeeaaaaaarrrrrr". He had the expression of a man sinking into a vat of warm marmalade while wearing nothing but his favourite socks.

"C'mon," I challenged him, "out with it!"

"I'm sorry, Fennel," said John, "but for most of the thirteen miles I was stuck behind a beautifully toned young lady with the most perfectly-formed derriere. She was squeezed into a tweed mini-skirt that nestled softly onto a well-sprung leather saddle that sent tiny little judders rippling up her buttocks. The whole perfect…peachy…package gyrated from side to side as she cycled along, never rising from the seat but, instead, grinding back and forth as she pumped her way up the inclines. It was the most erotic thing I've ever seen. I couldn't tell you where we've been, what gear I've been in, or how long we were cycling. In fact, I'm struggling to drink this champagne. You see, I have an irresistible urge to sink my teeth into a perfectly ripe…pear."

That's how I learnt that vintage cycling is the sexiest thing on two wheels, and getting high up on a Penny Farthing is the best way to get…a damn good look.

October

XIV

EXPERIENCES AT THE FLAT CAP CAFÉ

There are many Universal Facts of Life: things that are so tried and tested as to be irrefutable. These 'UFLs' range from the obvious, such as: 'Each of us will eventually grow old', to the obscure: 'and when we're old we'll feel compelled to wear fluffy-lined tartan slippers'. (Not to mention feeling irresistibly drawn to the sight of a certain fabric rubbing against warm leather.)

One UFL that gets my nod of approval (and absolutely no refuting whatsoever) is that cooking and eating food outdoors makes it taste infinitely better than the same meal prepared and consumed indoors.

Whilst gas-powered stoves provide the outdoor adventurer with instant and easily controllable heat, a wood fire is the master's choice. It adds extra flavour and is, to the connoisseur, more authentic. A man (always a man) who boastfully cooks over charcoal soon proclaims himself to be 'King of the Barbeque', even if he turns everything into blackened cinder. But the chef who can light a fire using means other than a match or lighter, and can then build that flame using foraged timber until it has a 'heart' hot enough to cook over?

Well, he or she has earned the right to wear the T-Shirt saying: 'Flame-grilled and Happy'.

I could go on, extolling the virtues of fire-steels over matches, but it's best not to theorise too much. As Agatha Christie said, "When engaged in eating, the brain should be the servant of the stomach". So let's 'keep things real' and get out there and start cooking.

My old friend Prof Winter described outdoor cooking as 'Eating at the Flat Cap Café'. By this he meant that men may not always be the cooks at home, but once they're outdoors they suddenly become masters of their field kitchens. A tin of soup? No problem. A boiled egg? Coming up. An eggy-cheesy-beans omelette on toast? You got it. But the ultimate meal to cook and eat outdoors? It has to be the 'Full English' fry-up of bacon, sausages, eggs, mushrooms, tomato, fried potatoes, black pudding and baked beans, all cooked together in a frying pan so large that it could double as the launch pad for an Apollo space rocket. That's what I call a meal. Good, honest, 'healthy' food that has absolutely no effect on one's cholesterol level or waistline. You can keep your dehydrated camp food. Just slap a dozen rashers of bacon on a pan and let fate do the rest.

There's much pride to be had in this 'one pot' or one pan cooking. Just how much food, and how extravagant a recipe, can be cooked in a single utensil? Does one throw all the ingredients in at once or add them in an ordered manner according to their individual cooking times? Are the sausages fat and juicy 'slow cookers' or thin and lean 'fast fryers'? Do you want crispy fried eggs or blubbery ones? And, once you've finished cooking, are you going to clean the pan or let next door's cat lick it clean?

Once I had the epicurean treat of sharing the 'ultimate cook-up' with a bunch of friends at the end of a week's fishing. We brought out our nicely festering week-old frying pans and threw in whatever food we had left, mashing it up for good measure. The resulting mess of sausages, bacon, sardines, kidney beans, digestive biscuits, brown sauce, luncheon meat, corn flakes and peanuts was bubbling along just fine until someone threw in a four-pound block of cheddar cheese. It melted down beautifully and gave the dish a lovely creamy texture, but it played havoc on our digestive systems afterwards. My friends Isaac and Angelus reported that they were unable to, as they put it, "release the beast" until four weeks after the meal. And even then it was, so they said, like "giving birth to ready-formed suet balls". (The learning from all this? Add several handfuls of raisins to the mix before eating, or replace one's toilet roll with coarse grade sandpaper.)

But it's not all lads' cooking and toilet humour. Recently, while on a fly-fishing adventure in the Welsh mountains with Thom Hunt, I had the jaw dropping and taste bud exploding delight of eating pan-fried trout fillets cooked in butter with shallots and new potatoes. It was Thom's way of thanking me for a super trip and just what you'd expect from one of the UK's leading experts on wild cookery. It was so simple and yet so delicious. I doubt if I'll ever taste food as good as that again. But of course, eating atop a mountain, with breath-taking views and great company, adds significantly to the experience. But the ultimate 'flat cap' challenge? It's about to happen. All thanks to my good mate Martin 'Northmore' Herrington.

Martin, if you don't know him, is an expert engineer whose day job is building aircraft for the RAF. He's also an angler and all-round countryman known for making fishing tackle and 'this and that' from whatever he can lay his hands on. Over the years he's sent me all sorts of things, including pictures of top quality nets and reels that he's casually 'knocked up' during his lunch hour. His latest gift, which arrived in today's post, is a five inch wide piece of tubular titanium with some holes drilled into it. Weighing next to nothing, yet being incredibly strong, I wondered whether it had fallen off the back of a Sea King helicopter?

Martin's letter, which accompanied the gift, gave the answer: "It's a hobo stove for your camping trips. Place it onto the base of your Kelly Kettle following a brew and you have an effective means of cooking your breakfast. Just keep adding twigs through the holes to maintain the fire."

EXPERIENCES AT THE FLAT CAP CAFÉ

Wow. In fact: wooooooooooooow. A new toy to play with. And one that could take centre stage at the Flat Cap Café. I'd have to try it out – and write about the experience in real time. So here goes…

I have pleasure in reporting that I am now sitting in the wood near to my home. I have my Kelly Kettle and hobo stove in front of me. To my left is a tray full of food waiting to be cooked. To my right is Mrs H's largest and most prized frying pan (a heavyweight brute-of-a-thing that, I imagine, would have a satisfyingly resonant Big Ben-like 'gong' should it ever be used to defend our property against burglars). Of course, I wouldn't normally bring a frying pan like this with me when camping. It's too heavy to carry comfortably in a rucksack and would take an age to heat up (plus, Mrs H would notice it was missing). But it's perfect for what I intend to do. You see, I want to find out if I can set a record for one pan cooking and perhaps begin a Fat Cap Café leader board for volume and extravagance of meals cooked over a hobo stove.

W. Somerset Maugham wrote, "To eat well in England you should have breakfast three times a day". So how about we cook all three meals at the same time, in one enormous pan?

Here's what I'll be cooking: 24 Cumberland sausages, 24 rashers of bacon, four rings of black pudding, six tomatoes, a punnet of mushrooms, two tins of baked beans, six eggs, a large tin of new potatoes, and two rounds of toast.

With the challenge set, I need to work out the tactics. Clearly, I need to begin cooking on a medium heat so to fry the sausages, bacon and potatoes slowly. Then I'll build more heat so that any water from the bacon evaporates and, when added, the fried eggs will have nice crispy edges. I'll need to arrange a stacking system for the food as it cooks (pushing it to the edge of the pan so to keep it warm while other ingredients cook in the centre over the heat). And perhaps I'll build a scaffolding system using willow skewers to hold it all in place? The baked beans would cook in their tins if placed alongside the hobo stove, and the toast could cook there too. So yes, I reckon I can do it. The only uncertainty is how to control the heat from the fire. What diameter twigs, at what frequency, should be added? There's only one way to find out: start the stopwatch with a "Ready, steady, cook!" and begin the Flat Cap Café chant:

> *"Sausage from a Cumber Land*
> *Rashers streaky, smoked by hand,*
> *Fruit from Tom who ate his toes*
> *Bread that's charred with fireside woes;*
> *Pudding black, a bloody treat!*
> *Potatoes new, for us to eat;*
> *Fungi grown on life departed*
> *Beans from Heinz whose wind once..."*

And there we are. All chanted, with food added, cooked, and ready to be eaten. A mountainous breakfast, piled atop one very proud frying pan. I guess you want to know how the stove performed? I couldn't possibly tell you that. In fact, I couldn't tell you much at all; at least until I've eaten the biggest and best breakfast I've ever seen.

I wish Martin were here to share the experience. Actually, I wish Martin could help me eat all this food. I've got a feeling there might be leftovers…

Stop – Unplug – Escape – Enjoy

What's your all-time favourite meal and most enjoyable way to cook?

November

XV

FOUR AND TWENTY BLACKBIRDS

It is said that cooking is the new rock 'n' roll. Celebrity chefs need only to be known by their first name, and a perfectly risen soufflé is more awesome than the sound of a Marshall amplifier turned up to eleven. Apparently.

Well, my name's Fennel. I rarely use my surname, and have proven my ability to cook a magnificent meal over a hobo stove. So, with the right apron and a sprinkling of salt, I could pass as a legitimate 'Jamie'. (Or, more likely, a total 'Delia'.) But knowing my luck, by the time I'd built a following, cooking would be out of fashion and unsuccessful cooks would use the chefs' names as expletives.

Encouraged by my recent success, I decided to bring my culinary excellence indoors. Mrs H finally had her new kitchen, so I had a brand new workspace in which to master my craft. It was gleamingly smart and full of all the mod cons my wife desired. She sang with joy while baking her first cakes in there, such was her newfound happiness. The new kitchen made up for all the years of sacrifice that she'd made while she saved up her money and put up with me. I could barely wait for her to be out of the house so I could use it to cook something 'ambitious'...

'Foodies', as cookery enthusiasts are known, often quote the 'local, seasonal, ethical' philosophies associated with good cooking and good eating. But my love of cooking is driven by something simpler than that: hunger. I'm a greedy so-and-so. I like good food (especially if I've hunted, foraged, or grown it myself) but not as much as the cosy feeling I get after a huge meal – when I slump to my armchair and savour the most contented of sleeps.

Give me a massive plate of wholesome grub any day. A pile of roast potatoes cooked in goose fat is more satisfying to me than a fancy meal served in a posh restaurant. Michelin Star? I'd rather chew on a French rubber tyre. Serve me all my daily calories on one plate, and make it as tasty as possible.

I've long been a fan of the concept of 'a meal in one'. Not the type served in a plastic pot to which you add boiling water, rather the sort that's wrapped in pastry, sealed, crimped and baked in an oven. I'm talking about the good old-fashioned English pie.

So-called "English" pies can be traced back to ancient Egypt. The Egyptians made primitive pies by mixing nuts, fruit and honey in bread dough. Drawings of such pies were etched onto the walls of the tomb of Rameses II, located in The Valley of the Kings. But pie pastry originated in Greece. The Greeks wrapped their meat in a paste of flour and water to seal in its juices. The Romans 'borrowed' the Greeks' recipe and transported it around Europe and ultimately to England.

Early medieval English pies, known as 'Coffins' or 'Coffyns', were savoury meat pies with tall straight-sided pastry crusts. The crusts, which were baked prior to being filled, had sealed bases. Once the contents and a pastry lid were added, they became effective cooking vessels. No washing up was needed, as diners ate the lot.

Typically nine inches in diameter, these pies were sometimes made much bigger so that they formed the centrepiece of the medieval feast. Being a fan of pies and big meals, I wondered whether I could recreate one of these super-sized medieval pies in Mrs H's new kitchen?

With Mrs H away at a conference, I was free to create 'the biggest and baddest' pie seen since King Henry IV of France's wedding day in 1600, when the guests sat down to eat their between-course entremets only to cut into the pie and see two dozen songbirds fly out.

With a pocketful of rye and sixpences aplenty, I set about finding a recipe that could be scaled-up to create something 'authentically dramatic': a heavy crust, high walled pie that would fill the oven to within an inch of its limit, which might become known as 'Fennel's Ultimate Pie'.

After finding a recipe in an old cookbook, I quickly rolled up my sleeves, raided the larder, and began assembling the biggest thing ever to emerge from a domestic kitchen. Here's what went into the pie:

The pastry*: built with inch thick walls, and pre-baked prior to being filled with the contents.
32 eggs, shells removed
16 tablespoons iced water
10 pounds of plain flour
5 pounds of butter
5 tablespoons salt
*Leave a large ball of pastry uncooked, for use as the lid once the contents have been added.

The contents:
10 pounds of stewing steak, diced, coated in flour, and fried to seal.
8 large onions, diced
48 medium mushrooms, whole
24 carrots, roughly chopped
12 sprouts, whole
4 parsnips, roughly chopped
8 medium potatoes, quartered
8 ounces of pearl barley

The gravy:
15 pints boiling water mixed with 16 stock cubes, poured over the pie's contents once added to the casing.

The resulting pie – with rather a lot of liquid leftover – measured 16 inches wide by 11 inches tall. It fitted neatly onto Mrs H's largest baking tray. It was a ground-thumper of a pie, plump of filling, warm of heart, and capable of satisfying many. Just what was needed to give me that contented feeling on a Sunday afternoon.

Ensuring a tight fit between lid and crust, and then making sure not to crack the sides, I lifted the pie into the hot oven. It fitted with barely an inch all round. I closed the door (just), and sat back, wondering how long I should leave it to cook?

A normal-sized pie would take thirty minutes to cook from cold. But this mega-pie was twenty times that size and contained raw ingredients. After much head scratching, I decided to apply the 'whopping-great-Christmas-turkey' timelines to the pie: twenty minutes per pound plus twenty minutes, all at 'one-eighty' degrees. I'd used approximately 38lbs of ingredients, so that gave me a cooking time of 13 hours. Hard-and-fast cooking rules being what they are, I closed the kitchen door and left the oven to do its thing. I retired to the living room, put my feet up on the sofa, and settled down to read all three volumes of Delia's *How to Cook*.

Twenty chapters in, and after doze in the chair, I was about to read 'Equipment for Serious Cooks' when I noticed a dry gagging sensation in my throat. I sat up in my chair and turned towards the kitchen.

There, at the base of the kitchen door, was greyish-black smoke rolling out across the carpet. I jumped forward, throwing the book on the floor, and ran towards the kitchen. Flinging the kitchen door open, I was hit by a wall of acrid smoke and a crackling sound.

Trying to calm the thumping in my chest, and wondering whether to call the fire brigade or the *Mary Berry Helpline*, I flung my arms around and opened a window. As the smoke cleared, I caught sight of the oven. Black bubbling gunk was oozing from its corners and dripping down onto the floor. The glass door was crusted with burnt pebbledash-like remains of pastry, meat and potato. Smoke was rising from its vent, and all the kitchen units were coated in sticky yellow condensation. I leapt towards the oven, turned it off and did my best not to choke on the smoke. I pulled at the oven door, which opened with a crunching sound as I prised it away from the oven's contents that looked and smelled like beefy charcoal.

The piecrust, it appeared, had set on fire and then developed a leak. All the gravy had escaped, dripping onto the oven floor where it had congealed into a tar-like mass before oozing free of the oven or bubbling into cinder. The pie, which had been sitting on the baking tray, had collapsed – filling the oven with its contents. The crust had then disintegrated, leaving a bird nest-like mess of meat and vegetables to slowly char and cremate.

I looked at the clock on the kitchen wall: 8.30am. The pie had been in for…what? Twenty-three hours? How long was that nap I had in the chair? Had Delia bored me into a coma? Oh, gawd, Mrs H would be back at lunchtime. That gave me four hours to clean up and remove any evidence of the 'pie-tastrophe'. Four hours? To scoop, scrub, mop, clean, air, and run? Oh Marco! Oh Gordon! Oh Nigella! What should I do? I was in a right Worrall-Thompson. Mrs H would have my Raymond Blancs for this. What a Blumenthal mess! A proper Fanny Craddock. Please, Ainsley-one but me?

Four hours. I'd better get started. Mrs H would be 'Ken Hom soon.

I had to focus, and work quickly. The kitchen had to be clean by the time Mrs H returned. I ran to the garden shed and fetched a bucket, a shovel, a broom, and a trowel. Sprinting back to the kitchen, I threw everything on the floor except the shovel. Holding it horizontally I plunged it into the hardened mess in the oven, scooping out what I could into the bucket. The centre of the pie was fine, but the bits sticking to the edge of the oven required some chipping off. So I got the trowel and scraped away at the walls of the oven until I could see the shiny metal coming through. Using a wet towel, I wiped out all the bits of leftover pie until the oven was empty. Then using some wire wool, again from the shed, I removed the final specks of black crusting from the oven walls.

Using the broom, I pushed all the gloop on the floor into a central slop, and then shovelled it into the bucket. But not having the foggiest idea where Mrs H kept her mop, and having no time to look, I ran to the garage and fetched the bucket and sponge I use to clean the car. I filled it with hot water and washing-up liquid, then thrust the sponge into it and splurged the liquid around the floor. It did a pretty good job, and soon the tiles were back to being their normal colour. Rinsing the sponge and filling the bucket again with fresh water, I continued the process of sponging down the kitchen by scrubbing the ceiling and then the kitchen units. It cut through the tacky substance that had coated everything and I was able to take a step back and assess the results. Not bad. In fact, pretty good. I couldn't remember the kitchen looking so shiny. The oven, which I turned on again to test that it was still working, positively dazzled. I carried all my cleaning kit outside, relaxing in the knowledge that I had an hour left before Mrs H would return. And when she did? There was no panic, no chastisement. Just a loving hug when she saw that I'd cleaned the kitchen.

"You wonderful star," she said. "You needn't have cleaned, as the kitchen's brand new; but it's such a lovely gesture that you did."

Sweet Fearnley-Whittingstalls, I'd got away with it. Or had I?

December

XVI

TWO PLUS FOUR

"Clothes make the man. Naked people have little or no influence in society."
Mark Twain

I doubt whether the male brain is, as we're led to believe, the result of evolution. Rather, I feel, it's the result of being dragged into the 21st Century by the hardworking and long suffering efforts of the opposite sex. Men are, after all, little more than chest-beating monkeys, whereas the so-called 'fairer' sex is usually in charge of the map, the fuel, *and* the engine. I was reminded of this today, when Mrs H accused me of having only a 'temporary interest' in cleaning the house. I'd apparently done a wonderful job of the kitchen when she was away, but had since shown 'no interest whatsoever' in washing dishes, mopping floors, dusting, hoovering, or peeling my socks off the bedroom carpet. It was as if my lack of compulsion to do these chores was a result of me being a useless man. How do I know? Because Mrs H gave me 'the look'.

In defence of my good wife, she was understandably

upset. Her new oven had developed a fault. Something to do with a strange burning smell it gets when it's been on for a while. She thinks the thermostat may be faulty, and has repeatedly asked me to get an engineer to come and fix it. Sadly, tried as I have, I've been 'unable' to find anyone who is prepared to inspect it. "It's just one of those things," I said to Mrs H. "These days, manufacturers expect you to return goods that are under warranty." I didn't have the heart to tell her that, when checking the warranty and viewing the oven on the manufacturer's website, I noticed that it's inner walls should be coated in black enamel. They were like this when we got it, but not any more. Best I didn't show the shiny metal walls of our oven to someone with a trained eye.

Anyway, back to 'the look'...

A man is biologically incapable of giving 'the look'. He might try, for example when his partner says that she has 'absolutely no desire to mow the lawn', but he cannot deliver it with the desired effect. She, on the other hand, can stare at him with brooding rage and make him feel as small and useless as the shrivelled-up droppings of a sand lizard.

I wonder what it is inside a man's brain that hardwires 'the look' into his guilty conscience? Think of John during his Tweed Run experience. He didn't *ask* to be stuck behind that tweed-skirted goddess. He didn't *choose* to think of her peachy bottom. He didn't *want* to watch it – for thirteen miles – gyrating on the saddle. He was just stuck there, unable to divert his gaze. Whereas a woman seeing the same skirt would probably only have thought, "Nice tweeds; I wonder where she bought them?" Of course, there's sexual attraction at play. But would a woman, when seeing a perfectly formed bottom (male or female) squeezed into tight tweeds, ever think "Nice arse?" She'd say she wouldn't, but I bet she would. Which makes me question whether it's not so much a difference between a man and woman's brain that's the issue, rather the communication trickery that goes on between the two sexes?

Time, I think, to conduct an experiment.

To begin with, I'm going to test your mathematical ability. Ready? Okay, here goes: If I had two beans in one hand and two beans in the other, how many beans would I have in total? Answer: four beans. Did you get it right? I hope so. (If you didn't, it's 2+2=4 and time for you to reach for a calculator.) Right, next one: if I had two beans in one hand and four in the other, how many would I have? Answer? Six.

This is where our experiment gets interesting. We are delving into the realms of male logic, which might enable a man – for the first time in history – to use 'the look'. (This could, given the theme of this book, be the most powerful Fine Thing he's ever possessed.)

Two plus four does indeed make six. Remember this, as we'll need to refer to it later. Also remember that

Fine Things that help us to define and communicate our identity. This allows me to explore a little side stream of thought while you're remembering the number six.

The way we clothe ourselves is possibly the best way to reveal our personality. Yet so many people feel obliged to follow the latest fashions. (As Quentin Crisp said, "Fashion is what you adopt when you don't know who you are". It supports the Greek philosopher Epictetus' advice to "Know first who you are, and then adorn yourself accordingly".) As a lifestyle author who also writes about the countryside, I wear traditionally styled country clothing at every opportunity. A flat cap, waistcoat, checked shirt, braces and breeks, and sturdy leather boots. That's my image. But I'm thinking of going one better. Actually, not one better: six better. *Still remembering the number six?*

American poet Richard Eberhart said: "Style is the perfection of a point of view". Therefore, if you have style then you must have perfected your point of view. My point of view is that we should have absolute confidence and pride in our self. Which has made me realise that my current choice of clothing is too 'rurally conventional'. I'm not a farmer; I'm a writer. So I'm going to upgrade my image by purchasing a *stupendous* pair of trousers. Something that a farmer would find impractical and only an eccentric with bold self-confidence could wear. Mrs H will undoubtedly give me 'the look' when she learns that I've bought them. But I'm committed. And, based on the success of our earlier experiment, I feel like chancing my luck.

To understand how the numbers will come to our aid, you need to know more about the trousers I'm going to purchase. So, do you know the difference between trousers, shorts, breeks, plus twos, and plus fours?

I'll explain: trousers reach the ankles, shorts end above the knees, breeks fasten immediately below the knees, plus twos fasten two inches below the knees, and plus fours hang four inches below the knees (they're cut eight inches below the knees, but are folded over and tucked into knee-high socks so that they balloon out and hang at the required distance). Understood? Good.

I'm going to purchase a pair of plus fours. And, as I'm likely to be wearing them a lot, I ought to purchase a second pair so that one can be washed while the other is worn. But, hang on a minute; didn't we calculate that two plus fours equals six? Oooh. That means I could, if I play the experiment right, purchase *six* pairs. Who could argue with that? Surely not Mrs H?

Well, as she's sitting next to me while I write this, we might as well ask her. So, in real time, here we go:

"Mrs H?"

"Yes." *Brilliant. She's replied. So far so good.*

"You're good with numbers, aren't you? After all, you do look after our money."

"Sure am, and I sure do. Do I need a calculator?"

"No, no. This should be easy for a maths whizz."

"Go on then, what do you need to know?"

"I need help with some adding up."

"Okay."

"It's tricky, so I'll say it slowly: If I had one bean…and then got another bean…and then fetched four more beans…how many beans would I have?"

"Is this a joke?"

"No. I'm being serious. I can work out the first bit, which is 'two beans'"

"So you need me to tell you the answer to 'two plus four?'"

"Er, yes please. Be a sweetheart would you?"

"Six."

"So you're telling me that two plus four is six?"

"Yes. Six beans. And six stupid-ass husbands who can't add for toffee."

"That's twelve."

"Oh, not so bad at maths now, are you?"

"Big numbers are easier. Besides, I just wanted you to confirm something for me."

"What?"

"That two plus fours is six"

"Yes, that's what I said."

"And you're happy with that?"

"Yes. The numbers do not lie."

"Good. So that means that if I were to buy a nice new pair of plus fours, and wanted to get a spare pair so that I can wear one pair while the other's being washed, then I really need to buy six pairs."

"Eh?"

"Hmmmmm," I think, with eyebrow raised. I'll give her 'the look'.

"What's that strange expression," she asks. "Are you having a wee?"

"What?"

"It's just that you looked like you'd lost control of your bodily functions. Six pairs did you say?"

"Yes. And I wasn't wetting myself. It's just that I was surprised that you couldn't see the obvious logic in the calculation: that two plus fours is six. Six pairs of brand new flamboyant tweedy trousers for me, that I can wear to show everyone how proud I am of my individuality."

"Oh, I see. So you're not *having* a piss then; you're taking it."

Oh no. Now she's giving me 'the look'.

"Well, you're always on at me about not ironing and suchlike," I plead. "This way, with me having six pairs of super-sexy legwear, you'll be under less pressure to wash them for me. You do realise that I'm only thinking of you."

"I see. And I understand. There is sense in what you're saying. And now you mention it, I wonder if you could help *me* with some maths?"

"Sure thing. Fire away babe."

"If I had ONE pair of shoes, then bought another ONE pair, then decided to return the ONE pair that I'd just purchased. How many pairs would I have?"

Right. Let me think. This ought to be easy.

"Erm, would the middle finger you're holding up have anything to do with the answer?"

Seems Mrs H knows me too well, and that I'll only be buying ONE pair of plus fours. Oh well. Was worth trying. Better luck next time...

December

XVII

TIP TO TOE

"If you're referring to my slightly unconventional mode of dress, I'll admit it must seem a little different... but to we bohemians it's quite normal, everyday, run-of-the-mill clobber."

Tony Hancock

Ralf Emerson said, "The greatest gift is a portion of thyself". This is true for those who are away from home for long periods of time, or have a spare kidney, but I believe that the greatest gifts are those that say, "I know you". How do I know? Because this afternoon, just before tea, I had my best soyou moment of the year.

If you don't know what a 'soyou' moment is, then I will explain. It's when someone comes up to you, points at something you're wearing or carrying, and says: "That's so you!"

To properly qualify for a souyou, the thing the person's referring to must have been given to you as a present. Why? Because it's difficult to be flattered like this when you've chosen the item for yourself. After all, where's the surprise in it? We know what suits us, right?

If you're lucky, the souyou comment will be about something you're proud to have about your person. For example, it might be a hat that someone gave you for Christmas which, when worn, says to everyone: "I was made for this head". If you're unlucky then it could be something you fear getting close to, like a Norwegian-knit reindeer pattern jumper with 'honkable' nose. The power of the soyou, therefore, depends upon how well someone knows you. If they give you a present that you absolutely hate, and someone still says, "That's so you!" then it's time to rethink how others perceive you. Or, in some cases, you might need to make an urgent trip to *Black Bags R Us*.

The worst soyou I ever received was in 1996 when, in my early twenties, my girlfriend at the time gave me one of those novelty baseball caps that holds two cans of lager and has little tubes from which you can sip the drinks that are sitting on your head. It was printed with the words: "Total Pisshead". I was forced to wear the present at her family's summer party, where I was introduced to most of her relatives. Three of them said, "Oh Fennel, that's sooo you!" It was like being dressed in the Emperor's New Clothes. Three negative soyous were enough to ensure swift disposal of both cap and girlfriend. But, as I was saying, today's event was a positive soyou. In fact, it was the best ever. To appreciate why, you need to know about the events of the past three weeks.

Three weeks ago, after Mrs H and I agreed that I would buy just one pair of plus fours, we received a formal-looking card in the post. Delivered by hand, it was printed with engraved lettering that read:

"Lady Archington-Smalls requests the pleasure of Mr & Mrs Hudson at her annual Cha and Chirping Party. The event, which supports local authors and artists, will be held on the afternoon of Saturday 1st December. Please dress appropriately. Nibbles will be provided. RSVP The Parsonage, Ham End."

"What on Earth is a Cha and Chirping Party?" I asked Mrs H.

"Not a clue," she replied. "There's a number on the back of the card; I'll give it a ring."

Mrs H picked up the phone and dialled the number, and I went to the kitchen to make a brew. While I was waiting for the kettle to boil, I could hear my beloved saying things like, "Oh, hmm, I see, ah, that makes sense, seems good, and it's in the garden you say, okay, we'll be there, see you then, oh, one more thing, why is it that if we 'dress appropriately' we will get nibbled?"

Mrs H ended the call and walked into the kitchen.

"It's a garden party," she said.

"What, in December?" I replied.

"Yes, but in a huge marquee. It's a networking event for creative people. Tea and canapés will be provided. Lady A-S knows of your books and thought you'd like to come along, to mingle a bit and give a reading."

"You're having a laugh?"

"Nope."

"I can't do that, they'll think I'm a nutter."

"Seems they already do, hence the request to *dress appropriately*."

"You know that sharp suits cause me to bleed, hence why I wear stinky old tweeds and hobnail boots. I can't wear my usual clothes to a posh event?"

"You'll need to find something. Cos we're going."

"But darling, people will have expectations. I'm no Lord of the Manor. I'm a writer. And we're skint!"

"Excuses, excuses."

"Seriously. I have absolutely *nothing* to wear."

"Okay. Leave it with me. We've yet to purchase your plus fours, so maybe we can include them in our plans. You focus on the networking, and I'll sort the rest."

Five days later, Mrs H presented me with small bundle of fabrics.

"What's this?" I asked.

"Samples," she replied.

"Er, babe, no. I'm not going down this route. The last time I looked at having a jacket made, it was going to cost five hundred pounds. It was way out of my league then, and even more so now. I've overspent this year, remember?"

"So you have. But Christmas is coming. And anyway, who does the finances around here?"

"You do."

"Exactly. So shut up and tell me that you like the third sample from the back."

I thumbed through the samples. They were all tweed fabrics, in colours ranging from a russet-fawn (which reminded me of Yorkshire moors in winter) to a moss-like dark green (which reminded me of a yew tree after rain). Rather pleasingly, each fabric was named after a game bird, which conjured up images of Edwardian shooting parties, fine cigars and large brandies. The first to catch my eye was the russet-fawn one, which was named 'Grouse'. This had a nice vintage look and I could see the individual hairs in the weave. The second was a darker tweed, a mixture of bottle and sedge green, with a burnt umber and red check pattern overlaid.

"Which one's that?" enquired Mrs H.

"It's called *Pheasant*. Rather nice, isn't it." I replied.

"Was it the third one from the back?"

"Yes. How did you know?"

"Because it's got your favourite burgundy colour running through it."

"Seems like you've already made up your mind."

"I have. In fact, here's the deal: we're going to the Lake District this weekend. I've arranged for you to meet Bob Parratt, the owner of the shop that sent me the samples. You'll try on a number of garments and, if they fit, I will buy them for you as your Christmas present. My only condition is that you stick to the

Pheasant tweed. That one's most 'you'."

"And you're sure we can afford it?"

"Trust me."

The weekend arrived and I found myself in one of my favourite landscapes in the world, visiting a shop filled with racks upon racks of tweed jackets, waistcoats, breeks, plus fours, coats, hats, jumpers and socks.

"What do you think?" enquired my beloved.

"It's great. I love all of it."

"Good. Then you may have one of each item, all in Pheasant tweed."

"What, a full shooting suit?"

"Yes. With a coat and shooting waistcoat as well. But no hat. You have enough of them already."

"What about the budget?"

"How much did you say that jacket cost ages ago?"

"Five hundred pounds."

"Well you can have all this for less than that."

"What? A three piece suit, with plus fours, a shooting waistcoat and a thermal coat? For the price of a tweed jacket? Really? That's amazing. I don't know how you do it?"

"Merry Christmas Mr H."

Which brings us up to today. At 3pm Mrs H and I drove along a gravel driveway to the Parsonage. We parked up and walked to the largest front door we'd ever seen. I stood there in my new suit feeling confident and relaxed.

TIP TO TOE

"Now remember what I told you," said my beloved.

"Yes," I said, "dress to the left and hope for the best."

"No. Don't be silly. Listen to my words: when it comes to being true to who you are, don't tiptoe about. Be confident. Be you. From tip to toe."

"Thanks babe."

"Oh, and don't forget to remove your cap before you enter. It's disrespectful for a man to wear a hat in someone else's home."

"Got it."

Knock. Knock. Knock.

We waited. We heard footsteps, and then the lifting of a latch. The door opened. A tall middle-aged lady greeted us and, in a remarkably posh accent, said, "Ahh, Darlings, you maaade it. Welcome, welcome! You must be Mrs H and oh, there he is, the man of the hour, Fennel, the countryside author. Do come in. And, oh, my, I just lovvvve the suit. It's soooo you!"

It was the best Christmas present I could have received. All thanks to my beautiful, long-suffering, and totally gorgeous wife: the person who knows me better than I know myself and who encourages me more than anyone else. She and the daughter we share are my most precious Fine Things. I love them, from tip to toe.

Stop – Unplug – Escape – Enjoy

Who are your Finest Things?
Are you as fine to them as they to you?

XVIII

LEGACY

I wonder what it will be like in the future when, in old age, we look back upon our lives? Will our memories be as bright as winter sunshine, or will they have faded into shapes shrouded in an autumn fog? Will we be able to look around our rooms and see all the treasured things that we've amassed over the years, and still remember their meaning? Will they tell us their tales, or will they have become dusty and anonymous? Will there be anyone around who's able to remind us why we kept these things? In caressing them, will they remind us of who we are and give us a heightened sense of being?

There comes a time in one's life, perhaps in middle age, when we stop and assess who we are, and the life we have. Studying the things around us is a good indicator of this. They're a fingerprint of who we are. Take a look at the things around you. What do you see? How do they look? Individually *and* collectively. Are they unique? Is the view that which your earlier self hoped to see? Is there anything you'd like to see, or not see, in the future?

I've done my best in this book to show you the things

that are important to me. Sure, there are things I've missed, but having a 'taste for life' is about knowing its various flavours and awaiting the next meal. Everyone will have a different list, which will change as we grow and develop. At times it will be a long list, carefully prioritised; at others it will be a shorter or more random collection. We could say that the fewer things on the list, the less needy we are. But I like to paint pictures. Our best canvas is all around us, in everything we touch and do. It pays to have a broad palette of things that define us.

Fine Things are reservoirs for the heart. They're souvenirs of life, a constant reminder of who we are and what we've done. Yet when our time comes and we prepare to meet our maker, we're unlikely to wish for more material things. "You can't take it with you when you're gone," they'll say, so the things become valuable in their ability to tell future generations about us. They become our legacy.

If you could choose just one thing as your legacy to the world, that most poignantly says who you are and what you stand for, what would it be? Would it be an object, a place, a song? Maybe a letter that you leave with your will? Or perhaps a book such as this? Only you can decide, and only they can decipher. That's the challenge, and it's one we'll all face at some time or other. But that is for them. What about us? We're more likely to wish, in our final hour, for more time.

LEGACY

More time to do the things we loved, with the people we loved, at the places we loved and to savour the Fine Things we loved. In this respect, life is the ultimate fine thing, with time the mortar that binds it together. As Charles Darwin said, "A man who dares to waste one hour of time has not discovered the value of life".

And so, as we approach the end of this book, I thank you for our time together. I trust that it's helped you to 'Stop – Unplug – Escape – Enjoy', and put you in good stead to embrace your most precious and ultimate fine things.

Life is a wonderfully fine thing. Go live it.

ABOUT THE AUTHOR

FENNEL HUDSON

"Author, artist, naturalist and countryman. His is a lifestyle to inspire the most bricked-up townie."

Fennel Hudson is a lifestyle and countryside author known for his *Fennel's Journal* books and as host of *The Contented Countryman* podcast. He's a naturalist and outdoorsman who, through his travels in wild places, explores the notions of freedom and self. A nature writer and country sportsman, his work has been compared to that of Roger Deakin, BB, and Jack Hargreaves.

Fennel observes the subtle things – around us and within us – that might otherwise go unnoticed. He values old-fashioned things and champions a relaxing and peaceful rural life. Much of his writing centres on what he calls 'The Quiet Fields' – those quiet corners of the landscape where time moves slowly and nature exists undisturbed. His motto, and the message in his writing, is 'Stop – Unplug – Escape – Enjoy.'

For more information please visit:
www.fennelspriory.com

THE FENNEL'S JOURNAL SERIES

THE FIRST-EVER REVIEWS OF FENNEL'S JOURNAL:

"Fennel's Journal began as a series of illustrated letters to friends. As these evolved they became less a diary, more a manifesto, and the Journal is now exactly that – a way of living, rurally and simply: very real for all those who recognise the importance of tradition and joy."

Caught by the River

"I can see where it might lead. What he has would make amazing TV. It's the Good Life, but in a realistic way. It's Jack Hargreaves. It's Countryfile. It's quality Sunday newspaper stuff. It's 1948, all over again. In trying to escape the present he's inevitably created a brand. A potentially very powerful brand."

Bob Roberts Online

"Fennel's Journal is a masterpiece about rural living. It is a route-map to the life we all seek."

The Traditional Fisherman's Forum

From A Meaningful Life:

"Life is the most beautiful and rewarding gift. We just need to take time out to allow us to reflect, change perspective, and see things in their best light. Sometimes we just have to stop and feel the pulse of the Earth, the rhythm of the seasons and the internal voice that was once our childhood friend. As the natural world grows smaller, so too does its intensity and the size of the window through which it may be viewed."

NO.1

A MEANINGFUL LIFE

A Meaningful Life is the first and perhaps most important Journal. It documents the origins of Fennel's Priory and why Fennel decided to live by a new set of ideals. With themes ranging from escapism, adventure, work-life balance, identity and purpose, through to traditionalism and country living, it sets the scene for future editions – building messages that are central to Fennel's Priory. Ultimately it conveys the importance of a relaxed, balanced, and meaningful life.

READER TESTIMONIALS

"I loved reading this Journal. It's inspiring and has the beginnings of something very special."

"Fennel's chosen trajectory is firmly in the slow lane. He's a countryman, with courage to stand behind his traditional values."

"Witty and emotive, Fennel's writing conveys passion for a slower-paced and quieter life."

From A Waterside Year:

"Water is intrinsically linked to the mystery and excitement of discovering new worlds. Of dreams. And hopes. And thoughts of what 'could be'. Dreams free us from normality. ...As the daydreams grew longer, the distinction between what was real and what was imaginary grew less. Soon I existed in a blissful world of my own creation. Reality, as I learned, is only a matter of perception...A life that is real to one is surreal to another."

NO. 2

A WATERSIDE YEAR

In *A Waterside Year*, Fennel takes time out to live beside a lake in rural England. Here he appreciates the healing qualities of water, studies the wildlife around him, lives at the pace of someone outside of normal daily life, and discovers the freedom that's found in isolation. Getting so close to Nature, and spending time in idle fashion, enables him to discover a stronger sense of self. Ultimately he learns that freedom is not a place, but something that exists within us.

READER TESTIMONIALS

"A year in the wild. How we would all love to follow in Fennel's stead and indulge our dreams, to come out the other side a stronger and wiser person."

"A Journal with a message – that we should take time out to think about what's important, and see the beauty of the world."

"A truly blissful read full of inspiration and humour. The story of Fennel sitting in his tent, with the noises outside, had me laughing out loud!"

From A Writer's Year:

"Writing, with a fountain pen and ink from a bottle, is the simplest of things. Yet it can transport us to a different place entirely. Imagination is the real magic that exists in this world. Look inwards, to see outwards. And capture it in writing."

NO. 3

A WRITER'S YEAR

A Writer's Year celebrates the writer's craft. It champions the handwritten letter, discusses vintage pens and writing ink, and celebrates things such as antique typewriters and the quirkiness of the creative mind. It's a blend of observations. It's funny. It's serious. It's real life. But most of all it is written to encourage aspiring authors to find their voice, to put pen to paper, and follow their dreams.

READER TESTIMONIALS

"Worth it for the first chapter alone. It cannot fail to motivate and inspire the would-be author."

"What Fennel has written is not so much a eulogy for the handwritten letter as a call-to-arms for everyone to follow their dreams and make the most of their God-given talents. This is a genuinely inspiring read."

"I loved the part: 'If a pen can communicate our thoughts, dreams and emotions and be the voice of our soul, then ink is the medium that carries the message'. It shows how important and generous writing can be."

From Wild Carp:

"Some will say that searching for your dreams is like looking for unicorns in an emerald forest. They will say that following a golden thread will lead only to a king, dethroned and living in the gutter. This may be so.
But the king was made, not born. The crown was never his to wear.
...If ever the adventure proves tiring, or you lose sight of your dream, look to the west at sunset. There, on days when the skies are clear, you might see upon the horizon a thin layer of golden mist. When it appears, you will know its purpose: it is the mist of believing."

NO. 4

WILD CARP

Angling for wild carp is about adventure, history, atmosphere and emotion. *Wild Carp* captures this aplenty, describing Fennel's 20-year quest to find a very special type of fish. But it's also about nature connection and a desire to uncover the seemingly impossible – a place where we can discover and live out our dreams, to completely indulge the mantra of 'Stop – Unplug – Escape – Enjoy'.

READER TESTIMONIALS

"When written well, traditional angling writing by the likes of BB, for example, is the type of literature that I can read again and again. Fennel's writing flows un-hurried without overly romanticising each point and the research is thorough; from the first sentence I was thinking, 'this lad can write!' It's informative and very refreshing."

"Such inspiring writing. His words 'Somewhere in the undergrowth of the impossible' had me staring out from the page in amazement. Fennel's writing is pure poetry."

From Fly Fishing:

"The deeper we travel into the natural world, and the greater the number of technological encumbrances we leave behind, the more likely we are to escape the fast-paced lifestyle and stresses of the 21st Century. For some, angling enables a quest into the unknown, an adventure into the wild. For these fortunate folk, fly-fishing is escapism. Their hours by water serve as contemplation to enrich their souls, directing their quest inwards, towards their longed-for state of completeness."

NO. 5

FLY FISHING

Fly Fishing celebrates the most graceful and artful form of angling, explaining what it means to be an angler – in the spirit of Izaak Walton – and how fly fishers differ from bait fishers. The sporting and aesthetic beauty of fly-fishing is described in Fennel's usual witty and contemplative style. As he says, "Fly fishing is the ultimate form of angling; it gives us a reason to fish simply, travel lightly, and explore wild places that replenish our soul. With a fly rod, we're not casting to a fish; rather to a circle of dreams: ripples that spread into every aspect of our lives".

READER TESTIMONIALS

"Brilliant writing. Fennel made me laugh out loud in bed. My wife was asking questions!"

"A delightful, well-articulated, read. I strongly recommend it, especially to the contemplative, tradition-loving, bamboo fly rod devotees among us."

"A very inspiring and rewarding read. I will try to tie the Sedgetastic fly. It looks tasty!"

From Traditional Angling:

"Physics teaches us that for every action, there is an equal and opposite reaction: a natural balance of energy that sustains the equilibrium of life. In modern angling, these forces are skewed so far in favour of technology that the balance between science and art has been lost. But there is a movement, an undercurrent that defies the flow of progress. There are those who choose not to follow the crowd. They seek not to fish in a predictable, scientific manner. They yearn for the opposite, to buck the trend, *to be different*. They are the Traditional Anglers."

NO. 6

TRADITIONAL ANGLING

Traditional Angling celebrates the Waltonian values of angling: about fishing in a seasonal and uncompetitive way for the pure pleasure of being beside water. It wears its heart on its sleeve and a wildflower in its lapel. It's passionate, provocative and eccentric, written for those who appreciate the aesthetics of angling and uphold its sporting traditions. So, with great enthusiasm, raise your bamboo rod aloft for an adventure that proves there's more to fishing than catching fish.

READER TESTIMONIALS

"A beautifully written, very engaging and hugely enjoyable read. In fact, it's the best thing on fishing I've read in a long time."

"What a Journal! Fennel is clearly the spiritual successor to his mentor – the great Bernard Venables. There's so much wisdom and craftsmanship in his writing. Bernard clearly taught him very well."

From The Quiet Fields:

"The countryside, with its vast horizons, fresh air and ever-changing seasons is, by its very nature, more life-giving and adventurous than any amount of modern indoor living. It inspires a love of natural history – everything from the birds that sing in the trees to the quality and richness of the soil beneath our feet. Most of all, it creates the desire to exist more naturally. And in doing so, we appreciate the balance of life."

NO. 7

THE QUIET FIELDS

The Quiet Fields is rooted in the humus-rich soil of the countryside. It's about remote rural places where Nature exists undisturbed, where we may sit and ponder 'The Wonder of the World'. The Journal tips its hat to these places, and to the nature writing of BB, revealing the 'Lost England' that still exists if you know where and how to look. It is the most sentimental and astutely observed Journal to date, discussing the 'true beauty' of Nature. If you've ever yearned to hear birdsong during a busy day, then this is the book for you.

READER TESTIMONIALS

"Fennel's writing reminds me of the works of Roger Deakin. It inspires me with faith in the quiet life and that although I may be isolated, I am certainly not alone."

"Fennel has captured the essence of the countryside – that is, its almost human character. So brilliantly has he compared and contrasted it with the nature of we humans. It's not so much a 'balanced study', more a 'study of the balance' between Nature and Man."

From Fine Things:

"It seems that, depending upon which side of the thesaurus-writer's gaze we sit, one's uniqueness as a person can be deemed to be either eccentric or distinctive. Both, in my opinion, are good...As we get older, and experience more things, those of us with strength of character and a sense of purpose will grow stronger and fight harder; those who lack identity and direction might end up sitting in a corner somewhere, blindly taking all the knocks that life throws at them. What does this teach us? That character and purpose are directly linked to confidence and conviction. What links them? Courage – to be oneself, no matter what others might say."

NO. 8

FINE THINGS

Fine Things celebrates the special and sentimental items and activities that convey our personality. The writing is fast-paced, quirky and humorous, reflecting the author's enthusiasm and eccentric view of the world. But be warned: if you look inside Fennel's mind, you might see a hula-hooping hamster named Gerald, shaking his maracas, loudly banging a bongo, and getting him into all sorts of trouble. So strap yourself in. This book picks up pace and takes some unexpected turns. From the deeply personal to the outright eccentric, it's for those who seek to be different.

READER TESTIMONIALS

"A very fine thing, indeed. Fennel's best and funniest book to date. He is the only author who can make me laugh out loud and cry in the same sentence. I was constantly in tears, for all the right reasons."

"Deep in places, outright bonkers in others. A demonstration of the fine line between genius and madness."

From A Gardener's Year:

"Roll up your sleeves and imagine your vision of paradise. This, in whatever form it takes, is your garden. Keep hold of the image; know it's every detail and piece together the elements that need creating or nurturing, so that when you get the chance, you can prepare the ground, sow the seeds, and make it real. Ours is a gardener's life, whether we realise it or not."

NO. 9

A GARDENER'S YEAR

A Gardener's Year celebrates the joy of growing things and reflects upon a life working with plants. But it's not a record of horticultural activities through the seasons. It's a metaphor for having a dream and making it come true. For Fennel, who has spent half his life working in gardens, it's about cultivating a cottage garden where he can aspire to a self-sufficient lifestyle. The Journal sees him sow the seeds of this future reality.

READER TESTIMONIALS

"Fennel's writing is uniquely funny. I mean, who else can name a chapter 'Chicken Poo'? His sense of humour, balanced with some deep yet subtle messages, had me in tears. From his 'escape' to a public toilet, to what not to say to a celebrity, this is a Journal to entertain all readers."

"When I started reading this Journal I had a garden with a lawn and a patio. Now I have a vegetable patch, blisters, an aching back, and the biggest smile of my life. Thank you Fennel!"

From The Lighter Side:

"If self-actualisation is the pinnacle of one's development, then it can't be achieved if your mountain has two peaks...Being the 'best version' of yourself implies that you have other versions kept locked in a closet. Don't have any 'versions'. Just have one true, beautiful and pure form of you.
So climb your mountain, open your arms to the Creator who greets you there, and sing loudly to the world that stretches out beneath you. Write your name permanently on the landscape of your mind. Remember: you are a child of Nature. And you are free."

NO. 10

THE LIGHTER SIDE

There's a delicate balance between something meaning a great deal and that same thing becoming so serious that it's ludicrous. (Ever got stressed about what clothes to wear for an interview?) That's why *The Lighter Side* provides the encouragement, humour, anecdotes, reflections and honesty that are essential to Fennel's message of 'Stop – Unplug – Escape – Enjoy'. After all, we can only 'Enjoy' if we know how to smile when we get there.

READER TESTIMONIALS

"The Lighter Side was more than I expected. The deeper meaning within it – and the devastating honesty it conveys – made me question exactly where I am in my own life and what I can do to improve it for my family and me in the time that remains. Thank you Fennel for opening my eyes and adjusting my course."

"The opening chapter is the most startling, erudite, compassionate and open piece of writing I have ever read...thank you Fennel for sharing so much. It did and does mean a great deal "

From Friendship:

"What I'm talking about is proper friendship. The sort that is authentic, genuine and real. Where we can look into the eyes of another person and know what they're thinking. ...Because, as friends, we remember 'why' as much as 'when' or 'what'. Through good times and bad, we were there. Together. That's the bond, the unquestionable obligation that's freely given. It's the tightest hug, the biggest kiss, the tearful hello and the widest smile. If that's what it means to be a friend, or an extrovert, or just someone who cares for others then that's me to the last beat of my heart."

NO. 11

FRIENDSHIP

Written by the Friends of the Priory, with bonus chapters from Fennel, *Friendship* provides insights into what it means to be friends, how shared interests and beliefs support collective purpose, and how, when we're together, we can achieve more, appreciate more, and have more fun. It's about the broader world of Fennel's Priory and how it exists in others. It's a book written 'for us by us', with friendship as the theme.

READER TESTIMONIALS

"Possibly the greatest gift that this Journal bestows is to let us know that we are not alone."

"Like friendship itself, this Journal brings together people and meaning. It reminds us that 'together we are strong'. Thank you Fennel for leading our charge."

"The message (and evolution) of Fennel's Journal is most evident in this Friendship edition. With such obvious themes as identity and legacy, it's clear that what Fennel has shared over the years is a route-map to freedom and a stronger sense of self."

From Nature Escape:

"I am once again seeking an escape, to where I hope to find freedom and connect with the young man who handed me his trust ten years ago. This will be a faithful interpretation of the Priory, a fitting way to mark ten years of writing. As I said at the end of last year's Journal, 'One's journey through life is not linear; it's circular.' So let's go back to the beginning, and rediscover the quiet world."

NO. 12

NATURE ESCAPE

Nature Escape provides the most detailed account of a day that follows the motto of 'Stop – Unplug – Escape – Enjoy'. In it Fennel returns to the woodland of his youth to study its wildlife and savour its peacefulness.

Written in real-time, with twenty-four chapters that each represent an hour, the Journal is an account of how time spent outdoors in wild places enables us to observe the nature that's around us *and* within us.

READER TESTIMONIALS

"Fennel's Journal has always provided us with an escape, but now we know where the escape can lead. As promised, it leads to enjoyment – and very enjoyable it is too!"

"24 hours alone in a wood, with only 'the wild' for company? With Fennel as our guide, there's no such thing as 'alone'; only the warmth of knowing that quiet times are the fine times."

"By studying the nature within us and around us, Fennel demonstrates how to be 'at one' with nature."

From Book of Secrets:

"There's a greater man than me who can sum up our journey, a mountaineer who in 1865 first climbed the Matterhorn. Edward Whymper, over to you: 'There have been joys too great to be described in words, and there have been griefs upon which I have not dared to dwell, and with these in mind I say, climb if you will, but remember that courage and strength are naught without prudence, and that a momentary negligence may destroy the happiness of a lifetime. Do nothing in haste, look well to each step, and from the beginning think what may be the end.'"

NO. 13

BOOK OF SECRETS

Book of Secrets links all editions of Fennel's Journal together. With 14 Journals in the series, and 14 core chapters in this book, it's the 'one book to bind them all' with each chapter providing the continuity story from one Journal to the next.

Containing Fennel's previously private writing, it provides deep insight into the Fennel's Journal story. If you've ever wondered why each Journal is themed the way it is, or tried to find the metaphor in each edition, then *Book of Secrets* is for you.

READER TESTIMONIALS

"What a privilege: being able to read the private writing of my favourite author. Book of Secrets is a treat."

"Such honesty and wit. Fennel puts into words what I have only ever thought, or dare not say."

"Fennel's Journal really is a series – it's meant to be read as a whole. And now we have the key to unlock it."

From The Pursuit of Life:

"We can hide, or we can strive – for a life of our making. With endless possibilities and opportunities to reach for our dreams, we owe it to ourselves to dream big and keep going, irrespective of what we might encounter. Sadly, the thing that most limits our success is not others, but ourselves. How strongly we believe, how confidently we act, how fiercely we react, how passionately we want, and how life-affirmingly compelled we are to grow and blossom; that's how we keep going, no matter what, to be the person we want to be, living the life we deserve, in dreams that are real."

NO. 14

THE PURSUIT OF LIFE

The Pursuit of Life concludes the Fennel's Journal story. It's a reflective tome that provides Fennel's commentary on the journey and a 'behind the scenes' view of the challenges and rewards of a life rebuilt on one's terms.

It's an account of how the series came to be and how it evolved, and includes much of Fennel's private writing, several of the original handwritten drafts, correspondence between The Friends, and encouragement for those on similar paths. Ultimately it shows how the Fennel's Journal series can be used as a route map to a more fulfilling life.

READER TESTIMONIALS

"A life retold, for our benefit. Fennel is to be congratulated for everything he's achieved – on paper and in life."

"It's his life in the books, but it could so very easily be ours. Fennel has a way of seeing truth in the severe and the sublime, and bringing it home."

"Can this really be the end? When dreams are real, we never wake from them. More books Fennel, please!"

www.ingramcontent.com/pod-product-compliance
Lightning Source LLC
Chambersburg PA
CBHW030331230426
43661CB00032B/1375/J